Wirt Robinson

A Flying Trip to the Tropics

A Record of an Ornithological Visit to the United States of Colombia...

Wirt Robinson

A Flying Trip to the Tropics
A Record of an Ornithological Visit to the United States of Colombia...

ISBN/EAN: 9783337143794

Printed in Europe, USA, Canada, Australia, Japan

Cover: Foto ©berggeist007 / pixelio.de

More available books at **www.hansebooks.com**

A FLYING TRIP TO THE TROPICS

A RECORD OF AN ORNITHOLOGICAL VISIT

TO THE

UNITED STATES OF COLOMBIA, SOUTH AMERICA
AND TO THE ISLAND OF CURAÇAO
WEST INDIES

IN THE YEAR 1892

BY

WIRT ROBINSON,

SECOND LIEUTENANT, FOURTH U. S. ARTILLERY

CAMBRIDGE
Printed at the Riverside Press
1895

PREFACE.

PREFACES are written for various purposes. Sometimes they are introductory, — they explain the *raison d'être* of the book, they define its scope, and perhaps outline the treatment of the subject; at other times they are self-laudatory, and impress upon the reader that the work fills a long-felt want, and that its statements are much more accurate than those of any other writer; again, in an humble tone they are apologetic, beseeching that the work be not harshly criticised. Should I make the last request in regard to the following work, I am afraid that I would be but calling attention to its failings. I am reminded of the story of the little boy who, visiting an art gallery where there was a statue bearing the placard, "Do not touch with canes or umbrellas," took out his pencil and added the words, " take a axe."

In a Preface it is proper and usual to return thanks to the various persons who have assisted the writer in the preparation of his work, and I should feel that I had been negligent in this respect did I fail to acknowledge the help that the officials of our National Museum have with great kindness extended to me.

It would be manifestly absurd should I attempt to criticise the countries that I visited, seeing what a short time I spent in them. My remarks must therefore be taken simply as observations of individual occurrences, not necessarily universal. I will say that throughout Colombia I met with a courteous treatment that we might sometimes look for in vain in many portions of our own country.

If I have dwelt too much on birds, remember that the study of birds is my hobby, that I went to the tropics for the purpose of observing them, and I am therefore inclined to give more prominence to them than to other objects. In my descriptions of them I have endeavored to give an idea of their approximate size by comparing them to some of our well-known birds. I have done this because I have often found that, from never having seen a specimen, I have had erroneous ideas of the size of some birds. Thus, I had thought from the figures that the larger hornbills were about the size of our crow, and, making an error in the opposite direction, I supposed that a stormy petrel was the size of a gull.

The illustrations, with a few exceptions which are noticed in the text, have been drawn expressly for this work or reproduced from my photographs.

CONTENTS.

LIST OF ILLUSTRATIONS.

x *LIST OF ILLUSTRATIONS.*

A FLYING TRIP TO THE TROPICS.

CHAPTER I.

THE VOYAGE.

I HAVE always been fond of Natural History in general, but especially of the study of birds, and at every new place that I have visited I have made it a point to look up the birds of the neighborhood on every opportunity, to study their habits and to acquaint myself with them as thoroughly as possible. As a result of this, it happens that I have seen a large percentage of our commoner birds of the Atlantic seaboard, and that, from familiarity with plates, drawings, and descriptions, I can recognize at once nearly every new one that I meet.

In the fall of 1891 I was stationed at the U. S. Military Academy. West Point, New York. My duties as instructor kept me occupied throughout the greater part of the week, but on Saturday afternoons I had a few hours which I usually devoted to rambling through the forests in the rear of the government reservation, on the lookout for whatever birds I might meet.

One such afternoon in November, I had returned from a long tramp over very rugged ground with a total of three species of birds observed : a pair of crows, a downy woodpecker, and a little band of six tomtits, — very meagre results for the seven or eight miles that I had gone over; and I was complaining about it to my wife. In the course of our conversation, I was led on to remark upon what I considered must be the enjoyment of a naturalist who finds himself for the first time in the tropics, surrounded by the most luxuriant vegetation ; where every object would be of the deepest interest to him; where every bird, animal, and insect that he should see would be new to him, and consequently afford him the same pleasure as if he had discovered it himself.   Imagine his delight when, after having fired at some bird

moving among the thick branches of a palm, he should pick up a trogon or humming-bird, brilliant with the colors of the most beautiful gems. At this point my wife said, "Well, why don't we go to the tropics some time?" and when we came to talk the matter over, there was really no unanswerable objection against our going; and so from that time we began to make plans for our trip.

My first act was to write to my brother Cabell, tell him of our project, and ask him to join us, to which he immediately replied that he would.

In selecting the point to be visited, there were a number of considerations that came up. First, our time would be limited; for which reason we should strike for the nearest point, so as to spend as little time as possible in going and coming. This indicated the West Indies or Central America; but our vacation would occur in June, July, and August, and these are rainy months in those regions. We could reach Venezuela in a little over six days from New York, but at that time that country was upset by civil war, and unsafe for travelers. To Panama the same objection applied as to Central America, and, in addition, there were vague rumors of yellow fever.

The interior of Colombia was found to answer our requirements, and was therefore selected as our destination.

During the winter we perfected our plans, got together our baggage, and tried to find out something about the country. This last proved to be a difficult task. I ransacked the various bookstores in New York, but nearly every book on Colombia that I found had been written during the twenties, and was therefore of but little assistance to us. I however found one that contained fairly good maps, and gave considerable information about roads, distances, etc. I refer to Holton's "New Granada."

In regard to our baggage: in the interior of the country it would have to be transported on the backs of mules, for which reason our trunks could not be larger than the ordinary steamer trunk, nor could they weigh over 125 pounds apiece, so that when they were slung, one on either side of the mule, the total load should not exceed 250 pounds. We were to carry two 12-gauge Parker's, one a very light smooth-bore, the other a heavy choke. My brother wrote that he would bring also his 32-calibre Winchester. I wished to carry paper shells, but economy of space made me decide upon brass ones, which could be reloaded an indefinite number of times. Our wads were taken from their boxes and put into shot-bags, as they could thus be packed more compactly. The powder we got in one-pound cans, and all of the above went into the trunks among our clothes. For shot, we took a good supply of dust, 8's, 3's, and a few buck, all done up in a stout bag that could be easily packed. For stuffing birds I carried a supply of arsenic, corn-meal, cotton, and scissors.

Wirt Robinson.

STEAMSHIP ROUTES TO COLOMBIA.

Upon looking into the matter, we found that there were three practical routes from New York to Savanilla (now Puerto Colombia), the seaport town for the interior of Colombia.   First, there was the Atlas Line, running to Savanilla, but touching at various ports in San Domingo, and thus stringing out the voyage to fourteen days ; secondly, the Pacific Mail Steamship Company's Line to Colon, on

THE VENEZUELA.

the Isthmus of Panama, in eight days, and from there three or four steamers per month to Savanilla ; and, thirdly, the Red " D " Line to Venezuelan ports, touching at the island of Curaçao on the sixth day out, and from this island various steamers of the English and German lines touching at Savanilla.   The greatest delay that we might have at Curaçao would be a week, whilst, on the other hand, we might make close connection, and for this reason we selected the last route.   We finally engaged our staterooms on the S. S. Venezuela, sailing Saturday, June 11.

I remember now with what feelings of delight I opened the letter from the steamship company, drew out the tags marked "passenger's baggage, S. S. Venezuela, Curaçao," and fastened them to our trunks. Our longed-for trip was finally assuming a tangible form.

It must not be supposed that our preparations progressed without opposition ; our friends all protested when they learned that we were going to South America in the *summer*. It was in vain that we represented to them that being so near the equator there would be but little difference in the temperature from one year's end to another. Our respective families and relations were disgusted. Their letters to us were filled with our obituaries, with stories of poisonous serpents, of all sorts of malignant and deadly fevers, of assassinations, and of lesser evils without end. I was reminded of the "Jumblies" in the nonsense book, who,

> "In spite of all their friends could say,
> In a sieve they went to sea."

Well, the winter and spring went by ; June 11 came at last, and found us together in New York. We left our hotel about eleven, drove down to Pier 36, East River, and went aboard the Venezuela about noon. We spent the time remaining before the sailing of our steamer in getting our luggage arranged in our very large and comfortable staterooms, and in examining the ship. The Venezuela was practically new, the staterooms very clean and well ventilated, the saloon and dining-room handsomely finished in quartered oak. She was of 2,300 tons, the largest vessel of the line.

The pilot came aboard a few minutes before one, and at one sharp we pulled out from the pier, headed down the bay, and started off. I had my "Hawk-eye" in readiness, and took parting shots at the Brooklyn Bridge and the Liberty Statue as we steamed by. The day was very pleasant and the sea smooth. When off Sandy Hook we slowed up, our pilot was taken off by his boat, and we started ahead again.

Shortly after this I saw my first stormy petrels. Quite a flock of them followed the steamer until it grew too dark to see. They were smaller than I expected to find them, — little gull-like birds with white rumps.

There are some people who laugh at seasickness, but I am unfortunately not among that number. In about an hour I began to feel wretched, and I grew steadily worse. Cabell also looked green. Alice held out better. When night came I would have been glad to die, and fell into my berth in a sort of stupor.

Let us not dwell upon a painful remembrance.

The following day, Sunday, June 12, when I crawled out on deck we were dashing through the Gulf Stream. I was at once struck by the change in the color of the water; it had now become of a most brilliant and beautiful dark blue, entirely different from the greenish blue of the water nearer the coast. Looking towards the stern of the vessel, I saw that we were still followed by a flock of the small petrels that I had seen the day before. They circled around the stern, every now and then dropping down to the foam in our wake to pick up some particle of food, and then hastening on to rejoin the retreating ship. They came within a few feet of the rail, and I, encouraged by a temporary lull in my symptoms, took my camera and went back to take a

"OUR PILOT WAS TAKEN OFF BY HIS BOAT."

snap shot at some of them, but the motion over the screw was so much greater than that amidships that I gave in before I succeeded, and re-treated to my stateroom more wretch-ed than ever. In the afternoon I saw a few flying-fish and some " Por-tuguese men-o'-war," the latter offer-ing a beautiful sight as they sailed lightly over the waves, resplendent with various shades of violet, purple, and pink.

Sunday night the wind freshened, and all day Monday we pitched through a head sea, the wind being from the southeast. We all felt worse than ever. I thought the sea very rough, as the waves repeat-edly washed over the decks. A flying-fish came on board and was caught. I exam-ined it as closely as I could. It was a small one, about six inches long, a deep blue color above and silvery white below, a splendid example of protective colora-tion, as its colors harmonized with the deep blue of the water and snowy white of the foam. Later in the day I saw a bird about the size of a pigeon, black above and white below, and more stocky than a tern. It flew close to the surface of the waves. It was not a tern, but flew much like a gull, not with the rapid wing-beats of a murre, and was probably a shearwater.

The wind continued on Tuesday, but not so fresh as on the day before, and, to

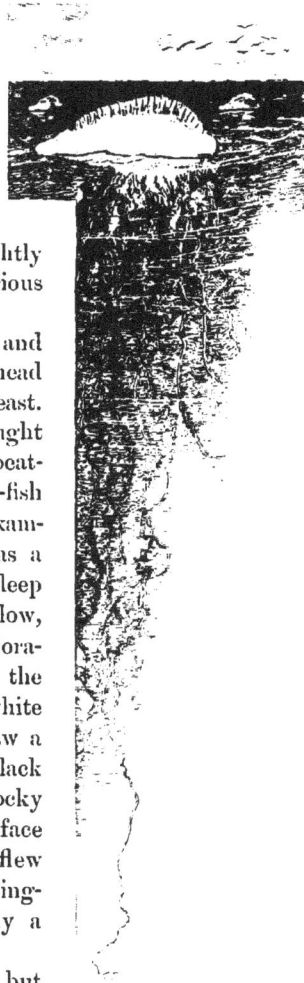

PORTUGUESE MAN-O'-WAR.
(By permission of American Book Co.)

my great relief, our symptoms began to moderate, and we once more took an interest in life. We began to make the acquaintance of the passengers, among whom we found especially agreeable Mr. Birtner, the German consul to Maracaibo, who was accompanied by his family. We also began to develop ravenous appetites and to look forward impatiently to meal-time, when we did full justice to the good dishes of the Venezuela's cook. Captain Hopkins was kind enough to place us at his table, and did a great deal to make our trip a pleasant one.

This day we saw hundreds of flying-fish, and watched a great many of them throughout their flight. They cannot be properly said to fly, yet they do more than simply sail through the air with the momentum acquired by their start from the water. As soon as they clear the water, they spread their wide pectoral and anal fins and hold them horizontal and motionless during the remainder of their flight. They can steer themselves up or down, as I saw hundreds of them keep at a distance of a few inches above the surface, going down into the troughs between the waves, but rising to clear the crests. I also saw some, when they had lost most of their velocity and were apparently just about to return to the water, droop the hinder part of their bodies until their tails touched the water, when they would wriggle them rapidly and violently and thus get a new impetus without actually entering the water.

Wednesday was like Tuesday; the wind was still against us, so we did not go along as rapidly as we otherwise would; still we averaged about three hundred miles per day.

On Thursday morning, as we made the Mona passage, we saw our first land since leaving New York: Mona rock, a sharp and rugged peak rising from the water on our left; Mona Island, a large, barren-looking table-land, with precipitous and, in some places, overhanging shores on our right. To the extreme right was a little flat sand-bar of an island, Little Mona, or Monita, and in the far distance to the left rose the blue mountains of Puerto Rico.

As we drew near the passage, many birds came around the ship;

among them a beautiful tropic bird with a yellow beak, white plumage with black wing patches, and long white plumes in its tail, sooty and noddy terns, flocks of boobies, the adults brown above and white below, the young uniform plain brown with bluish green faces and beaks. These attempted several times to settle on the rigging

YELLOW-BILLED TROPIC BIRD (PHAËTHON FLAVIROSTRIS).

of the vessel. Late in the afternoon we saw a tremendous school of porpoises — all small ones not over four feet long, but there were at least three hundred of them.

The weather was pleasant ; there were a few showers, but the trade wind was constant, and we did not feel the heat.

CHAPTER II.

THE ISLAND OF CURAÇAO.

FRIDAY, June 17, 1892. We were all up bright and early, keeping a sharp lookout for the first sight of land, and about eleven o'clock we saw away off on the horizon a faint blue peak which rose from the sea, as we drew nearer, and finally spread out into the Island of Curaçao. This island, which now belongs to the Dutch, is long and narrow, and lies with its longer axis N. W. and S. E. It is forty miles long and about ten miles wide, and contains an area of two hundred and twelve square miles. It is about fifty miles from the mainland of South America, and as we rounded its northwest end, and ran down its southwest coast, we saw to our right a faint blue line of mountains, the peninsula of Coro. We steamed along at a distance of from two to three miles from the coast for about twenty-five miles, and had a good opportunity to examine the island. It consists of a succession of peaks, some with a gradual slope on one side and abrupt cliff on the other, others with a flat top and abrupt fall on all sides. It is of coral formation, and all along the coast of the northern part there are miniature cliffs of from ten to twenty feet high, and at their feet long stretches of most dazzlingly clean-looking sand. The waves have undermined these cliffs in a number of places, making small caves along the water's edge. I have never seen a more beautiful sight than the deep blue waters of the Caribbean Sea breaking in waves on the smooth beaches of Curaçao. As the water grew shallower, the blue changed in shade to the color called peacock-blue, and this closer in became a light green.

VAR 2° E.

Savonet

Mt. ST. CHRISTOFFEL
TAFELBERG

MT ST. ANTONIE.

ST. KRUIS BAY

HATO

BEACH

ST MARTHAS BAY

ST JANS BAY

AT.

BULLEN BAY

CAPE ST. MARIE

ZWAK

SCALE $\frac{1}{200.000}$.

0   1   2   3   4   5   6   7   8   9   10 MILES.

MAP OF THE
ISLAND OF CURAÇAO
DUTCH WEST INDIES
FROM OFFICIAL SURVEYS OF
1888.

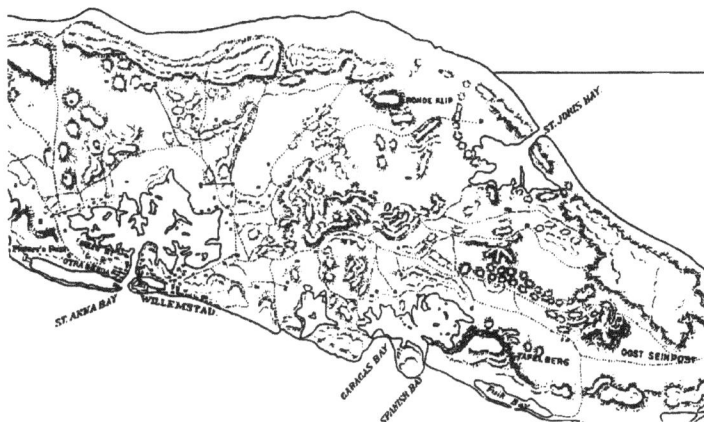

From a distance the island looked green ; but as we drew nearer,
it was seen that the greater part was practically barren.   The coral
rock showed everywhere, and was covered with a small scrubby
growth hardly waist high.   In the valleys between the peaks were
a few trees.   Farther south the shore grew more level, the beaches
wider, and at one place there was a large mangrove swamp.

Shortly after one o'clock we knew that we had been sighted, for
we saw the signal flying from a staff on one of the peaks to the
northwest of the harbor of Santa Ana, and later the little town of

VESSEL PASSING BETWEEN FORTS AT NARROW ENTRANCE OF HARBOR OF CURAÇAO.
(From Photograph by Ugueto.)

Willemstad came into view, the houses looking so charmingly neat
and fresh colored that they seemed to be china toys.

About half past two we were outside of the town ; we drew nearer
the shore, steamed slowly along past the entrance to the harbor,
picked up the venerable-looking white-haired pilot who came out to

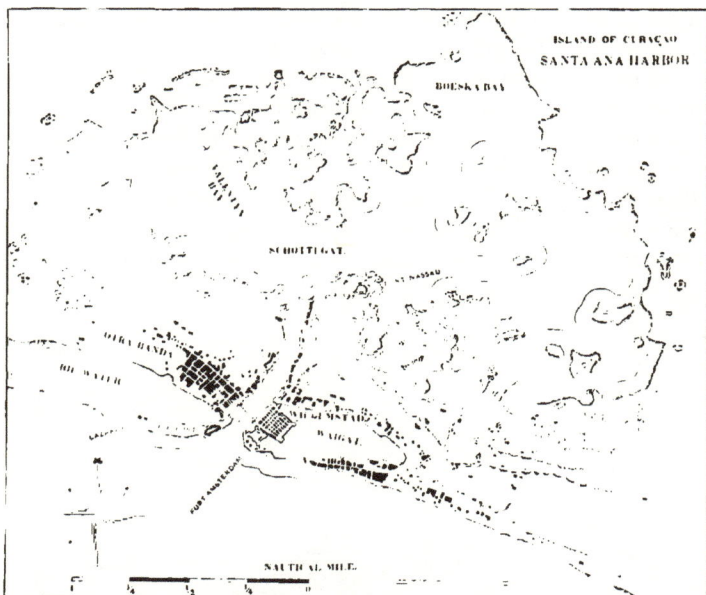

ISLAND OF CURAÇAO
SANTA ANA HARBOR

us in a whale-boat pulled by four very black negroes, then wheeled sharply around to our right, and continued on the circle until it brought us in between the two forts guarding the entrance.

This is a very strange harbor; its entrance nearly perpendicular to the coast-line, hardly one hundred yards wide, and continuing inland, more like a canal than anything else, for nearly a mile, when it opens out into a large and very irregular bay called the Schotte-gat, or more generally the Lagoon. This canal is about one hundred and seventy-five yards wide at its widest part, yet runs from forty-five to ninety feet in depth, with its shores so steep that the largest steamers can safely make fast within a few feet of the pave-ments, and at some points actually tie up alongside. There are no streams on the island, no running water, and no current or tide in this harbor. Its formation is due to the coral structure of the island.

The town lies on both sides of the harbor mouth, but does not extend back to the Lagoon. The portion to the right is called Willemstad, whilst that to the left is called Overzijde or Otrabanda, which are Dutch and Spanish respectively, and mean about the same as the English "other shore." Just before the channel debouches into the Lagoon, the land on either side rises into rugged hills, the one to the right being the higher and being surmounted by a small fortification, Fort Nassau. Owing to the elevation of this fort, it commands a view of the sea for a long distance around, and from it are displayed signals announcing the approach of vessels. From it, also, a time-gun is fired daily.

After passing the forts with groups of Dutch soldiers in curious ill-fitting uniforms, ridiculously tall forage caps, and short heavy swords at their sides, we went through a drawbridge of rather novel construction, proceeded a quarter of a mile inland, and finally our vessel turned around (though there hardly seemed room for it to do so), and we tied up along the western shore, sparred off to a distance of fifteen feet. The water is wonderfully clear, and we saw numbers of

DUTCH SOLDIER AT CURAÇAO.

fish of different kinds and sizes swimming about. There were several other steamers in the harbor, the Caracas of the Red "D" Line bound north, the branch steamer Maracaibo, a German steamer, and beyond, in the Lagoon, a couple of small men-of-war, Spanish and Dutch.

Our vessel was soon surrounded by small boats, flat-bottomed, square at each end, sculled by very large and very black negroes who stood on the back seat. (See illustration on page 13.) They brought out the port officer and runners from the hotels across the harbor from us.

Tired of being cooped up on shipboard, we thought of going over to spend the night at one of the hotels, although Captain Hopkins was kind enough to ask us to remain on the Venezuela. However, as I had some misgivings, I concluded to leave our baggage on board until we had made an inspection, so we took one of the small boats and went across, first to the Hotel Commercio, where we were shown up a flight of steep and rickety stairs to some whitewashed, bare, and unattractive rooms over a store; then we went to the Hotel Sasso, which we found worse, and finally, discouraged by the outlook, we concluded to accept the captain's invitation. Everything is comparative in this world. In less than two months we were delighted to get rooms at the Commercio, and found them extremely comfortable.

After this we took a short walk through the streets. We saw swarms of negroes in every direction, men and women, both remarkable for their fine size. The men wore straw hats, a light shirt, a pair of trousers, and were barefooted. The women wore turbans, one dress, and were barefooted, or at best wore slipshod slippers or alpargatas. Some of them wore dresses but little below the knee, others had long stiff-starched trains scraping and rattling over the pavement behind them, whilst the front of the dress cleared the ground by a foot. Children went naked, or wore but one ragged garment. We saw one boy of eight or nine with nothing but an old buttonless waistcoat which had belonged to a stout man, and which flapped around his knees.

The women carried their children astride of one hip; everything else they carried balanced on the head. We soon found it so hot that we returned to the steamer, and later Cabell and myself went out for a walk, leaving Alice on board. We strolled around the streets for about an hour, and then came back. We crossed the drawbridge through which we had passed earlier in the day. It is a pontoon bridge, a number of whose centre spans are fastened rigidly together by the road-bed, so that the whole swings open like a gate. On the pontoon farthest from the pivot is a donkey-engine such as is used on shipboard. To open the bridge, this engine takes in a rope fastened to an anchor up-stream; to close it, it hauls in on a rope in the opposite direction. It is a toll-bridge, the toll

DRAWBRIDGE ACROSS THE HARBOR AT CURAÇAO.
(From Photograph by Soublette.)

being two coppers of Dutch money, about equivalent to one cent in our currency.

The town is very picturesque; the houses and streets are remarkably neat looking. Though the island is so near the mainland, where it rains frequently, here it rains but seldom; sometimes two years go by without rain. There are no springs or good wells, and for drinking water cisterns are depended upon. There is, strange to

say, little or no dust.   The houses are of stone covered with stucco
or plastered, and are painted or washed in different colors.   Yellow
is the prevailing color, but a number are blue, green, white, and
pink.   The roofs are covered with red tiles.   Few of the windows

DWELLINGS AT CURAÇAO.

are glazed, although all have heavy blinds, usually green and
white, and the lower windows have large iron or wooden bars built
in.   The trimmings, door-frames, house corners, and ridges of the
roofs are painted white.   A great many of the houses have their
gable ends facing the street, but the slope of the end walls is
prettily broken into curves and angles, with appropriate moulding
all along.   There are no chimneys to the houses.   Cooking is done
over a handful of twigs or charcoal in a little iron or earthenware
vase like a fruit-dish.   They are much like a plumber's stove, or

like the stoves that our laundresses use to heat their irons. Of course one is required for each dish, as only one article at the time can be cooked on them. The stores are well supplied, and as this is a free port everything is extremely cheap, — many things being much cheaper than they are in the United States.

The streets, some of which are too narrow for any vehicle, have no sidewalks, but are all neatly paved with water-worn coral blocks set in mortar. The pavements are put down in regular pattern, square sections with diagonal lines, like the letter X. We saw a

NARROW STREET IN CURAÇAO.

funny little street car drawn by a donkey. There were seats for only six passengers, and the car carried a driver and a conductor.

In the shade in front of houses, and in a great many doorways, squatted old negresses with fruits, peanuts, candies, dried fish, and charcoal for sale. Among the tropical fruits which I tasted for the first time were some "mamones," a fruit which grows in bunches and looks like a large green grape. The skin was rougher and thicker, and when bitten split open, showing a sweetish, yellowish pulp around a large stone. I also tasted some mangoes, a large pear-shaped fruit with a smooth yellowish green skin. This, when peeled off, showed a soft yellow pulp, something like our pawpaw but more fibrous. It had a sicky sweet taste, with a flavor of turpentine that made it very disagreeable to me. I also saw a fruit called "cachú," pear-shaped, pink and yellow, with a lead-colored bean-shaped excrescence at the larger end. The fruit which they speak of highly here, the "níspero," we did not get.

The different kinds of money in circulation here is remarkable: old Spanish, Portuguese, Venezuelan, English, Dutch, French, — in fact, all kinds of coin. It is rather confusing to attempt to pay an account made out in guilders and florins from change consisting of francs, shillings, and reals. American gold, silver, and paper passes freely, but not the five-cent nickel. Speaking of money reminds me that an American contemplating a visit to South America need never trouble to get English gold. American gold is taken, and passes freely everywhere.

The natives speak a mongrel dialect called "Papamiento," and even have several papers printed in it. It resembles Spanish somewhat, but includes a number of words of Dutch derivation. I found it almost unintelligible. I succeeded, after a fashion, in making myself understood in Spanish, as nearly all of the natives speak a little of that language.

Of domestic animals we saw a few small horses, donkeys about waist high, curs, goats, sheep, chickens, turkeys, pigeons, and muscovy ducks. We saw for sale at different places a number of young parrakeets, green, with dirty yellow or buff-colored heads (*Conurus pertinax*). They were not fully feathered, and we were told that they had been taken towards the northwestern end of the island.

In a negro's house, near the steamer, we saw in a cage a number of young birds, none of them fully fledged. There were some doves, which were the same as the little ground dove of our Southern States (*Columbigallina passerina*). There was also a pigeon, considerably larger, of a wine-colored gray, with white feathers in its wings (*Columba gymnopthalma*). This was an undeveloped squab. The man called it "paloma con alas blancas," white-winged dove. There were also three partridges, which, at first sight, I thought were the same as our Virginia bob-white; but I soon saw that they were different. They were about half grown, and had a marked resemblance to the bob-white in shape and in coloration of their backs and tails. Their throats were white, with some reddish brown feathers among them; but the distinguishing feature was a long recurved crest of whitish feathers, which they carried continually erect (*Eupsychortyx cristatus*). All of these had been caught within a short distance of the town. We also saw flying about, and heard it singing, a bird very much like our mocking-bird (*Mimus gilvus rostratus*). I saw hovering over some flowers on the parapet of one of the forts a small brilliantly green humming-bird (*Chlorostilbon atala*).

On the stones in the water's edge along the harbor we saw quantities of sea-urchins, with spines eight inches long, barred with black and white (*Diadema setosum*). Before turning in for the night, we decided to go out with a gun early the next morning, and I arranged for the negro who had the caged birds to go along with us as a guide.

Saturday, June 18, 1892. I was awake by five o'clock; woke Cabell, and we dressed hurriedly, and left the ship, taking our smaller gun and only fifteen squibs of dust-shot and a few heavy cartridges. We found our guide waiting patiently for us, and struck off up the hill to the northwest. It is forbidden for any one to go through the streets with a gun here; but I had on a hunting coat, with voluminous game pockets, in which I put the stock and barrels, and did not put my gun together until we were beyond the limits of the town. We had hardly gone two hundred yards when we began to see numbers of the small ground doves; and I shot two, one a male in fine plumage. They were, as I had thought, the same as those found in our Southern States.

We first followed out a ridge running west for about a mile and a half. The country was very rough and hilly, the rock outcropping in every direction. In places, the ground was covered with fragments of the fossil coral, looking like pieces of bones; in others, the outcropping rock was as rugged and sharp as slag from a blast furnace. The surface was covered with a dry, thorny scrub, about three feet high, and the stems of this scrub were loaded with small, oblong, oval snails, about the size and shape of a 32-calibre rifle-ball. In walking along we crushed so many under foot that our shoes were made quite sticky. In this scrub I saw and heard several little yellow birds, and shot one, which, on picking up, I found to be a warbler, — a male. It was much like our yellow warbler, except that its forehead and crown were chestnut (*Dendroica*

CURAÇAO LAND SHELL. (PUPA UVA, LINN.).

rufopileata). Along here we also saw a number of small finch-like birds, and Cabell shot a pair (*Euetheia bicolor*). The male was dark slate about the head and breast, the rest of its plumage greenish gray. The female was plain greenish gray. They have very high culmens, and look like little grosbeaks. We heard them singing in all directions. Farther on we turned to our right, and went down into a little valley, where there was a small pool of brackish water, and here were some few trees, a couple of tamarinds, some date palms, and a number of calabash-trees. The calabashes are spherical or oval, smooth, and green like small watermelons, and grow from the trunk of the tree or side of the large

limbs, and not at the end of a twig. We ate some of the tamarinds, and found them quite refreshing. There was also another
scrubby tree, hardly fit to be called a tree, with straggling thorny
limbs and small leaves, like our honey locust. This tree was scattered pretty generally over the hills, and we noticed a peculiarity
about it, that is, that the majority of its branches pointed towards
the west. This is a result of the trade wind blowing constantly
from the east. This tree bore a few tiny yellow blossoms, and
around these we found some humming-birds. I missed the first
one that I shot at; but later Cabell killed a pair. They were
smaller than our ruby-throat, the male a
most beautiful glittering green, its tail
steely blue, almost black, its wings dark
purplish brown. The second was either a
female, or young, and was similar to the
first, except that its colors were less brilliant. It had some dark grayish feathers
below and a white streak on each side of its
head. Both had little downy white puffs
around their vents (*Chlorostilbon atala*).

CHLOROSTILBON SPLENDIDUS.
(After Elliot.)

A little later I shot a large sparrow, quite like our white-crowned
sparrow. Its head was handsomely marked with black and gray,
and it had a chestnut collar at the back of its neck (*Zonotrichia
pileata*). In a calabash-tree near here I shot a species of honey-
creeper (*Coereba uropygialis*). It was slate-brown above, its breast
and rump yellow, its head and throat slate-black, with a white
stripe above each eye. There was a fleshy excrescence at its gape,
which was pinkish red when the bird was fresh, but which faded
rapidly. Its tongue had a peculiar brushy tip.

We went on as far as a convent and an orphan asylum, where one
of the nuns, a negress, gave us a drink of water. We then turned
back, and reached the ship about nine. The roads near the convent
were excellent, and had on either side a hedge of a species of cactus
which grew up like tall posts to ten, fifteen, and even twenty feet in

height. We found other kinds growing about; one especially troublesome resembled our prickly pear, but had very long thorns. These appeared to have barbs on them, for when they entered the flesh they had to be picked to pieces before they could be extracted.

"CACTUS . . . TEN, FIFTEEN, AND EVEN TWENTY FEET IN HEIGHT."

When we brushed against one of these plants, a whole segment would break off and hang dangling from our clothes. On our way back we saw a pair of small hawks (*Tinnunculus sparverius brevipennis*), and got a couple of good shots at them, but the cartridges that I happened to have with me had been loaded for several years and were worthless, so we failed to get one. They seemed to be much the same as our sparrow-hawk. Our guide said that they were called "chiki chiki," from their cry, which much resembled this word, and that they fed on the lizards, "larguitos," which literally swarmed through the scrub, — repulsive-looking creatures, some green, some brown, and all spotted and blotched with lighter color. I was told that the green ones were males. They lived in burrows.

We also saw at a distance some yellow and black orioles (*Icterus xanthornus curasoënsis*). Our guide called them "tropiales," but they were not the common troupial. We saw numbers of the mocking-birds, but I had no more cartridges, so could not get any. The guide called them " ruiseñor," which is Spanish for nightingale. I saw a small red butterfly, and some very small grayish blue ones.

When we returned, we found that in our absence an English

MAIN STREET, CURAÇAO.
(From Photograph by Soublette.)

tramp steamer, the Navigator, of the Harrison Line, had come in and would sail for Savanilla the same afternoon, so I hastened over to see her commander, Captain Owen, and secure our staterooms. I found the Navigator to be a large freight steamer with only six staterooms, the accommodations being naturally far inferior in

every respect to those of the Venezuela. However, we were anxious
to hurry on, so I took passage for us, for which I had to pay twenty
dollars apiece in gold. We took lunch on the Venezuela, and afterwards I skinned the birds that
we had shot in the morning. I
had never skinned a humming-
bird before, and the first one that
I tried was such a sorry-looking

object when I had fin-
ished, that I simply
opened the second, took
out the intestines, and filled it with dry arsenic. This is the way
that I preserved nearly all of the humming-birds that I secured on
this trip, and I afterwards had cause to regret it. Though they
look well enough at first, and though the flesh is preserved, it
shrivels until the skin is distorted ; and, again, if the birds are
packed away in a trunk for a week or ten days without being
aired, they are apt to be mouldy and mildewed when taken out. I
should advise all collectors to skin their humming-birds as they do
larger birds.

Later in the afternoon we took a short walk through the streets,
went into the old Dutch fort to the post-office, mailed some letters,
came back to say good-by to our friends on the Venezuela, then
had our baggage taken over to the Navigator, and settled ourselves
in our staterooms. As we crossed the harbor, I saw flying over,
high in the air, a frigate pelican (*Fregata aquila*). It sailed along
gracefully, opening and closing its scissor-like tail.

We cast loose our lines about half past four, soon passed out of

the harbor, headed slightly to the north of west, and before nightfall Curaçao had sunk out of our sight. We ran along with a strong current and wind in our favor, and the ship rolled considerably, but fortunately I had no recurrence of my unpleasant experience on the Venezuela.

The second night before reaching Curaçao, I saw for the first time that constellation of which we have all so often heard, the Southern Cross, and this night we had a much better view of it. I must confess to being greatly disappointed. The stars are not so bright as I had been taught to expect, nor is the cross a symmetrical one in any way. The arms are not perpendicular to the vertical part, nor are they of equal length. The accompanying figure

THE SOUTHERN CROSS OF THE GEOGRAPHIES AND THE TRUE SOUTHERN CROSS.

gives the Southern Cross as represented in the geography that I studied when a schoolboy, and a second figure of the cross in more nearly its true proportions.

We were on board all the day of Sunday, June 19, and went along nicely with wind and current in our favor. I saw during the day a few petrels, and some large gannets, white, with black wing-tips, like the common one of our north Atlantic coast. We expected to reach Savanilla on the following morning. It was cloudy and hot during the day, and there were several small showers.

This would seem to be an appropriate place to make a few remarks about Colombia. I will not attempt to give a lengthy account of the country; for this I would refer to the Encyclopædia, to Bulletin No. 33 of the Bureau of American Republics, or to some of the works mentioned in the appendix; but I will simply refer to some of the leading features.

The Republic of Colombia consists of nine divisions or departments, each having a capital of its own, and is situated in the northwest corner of South America. Its northwestern extremity, the department of Panama, joins Central America; on the southern boundary is the Republic of Ecuador, and to the east lies Venezuela and Brazil. Our ideas of the relative size of the South American republics are apt to be vague. For instance, the area of Colombia is over 500,000 square miles, or equal to the combined area of the New England States, New York, New Jersey, Pennsylvania, Delaware, Ohio, Indiana, Virginia, West Virginia, Kentucky, Tennessee, North Carolina, and Georgia. It is of irregular shape; its greatest length is about 1,250 miles, its breadth 1,100 miles, that is, each dimension is, roughly, a third greater than the distance from New York to Chicago. It is one of the most mountainous countries in the world. The great Andes of Ecuador, crossing its southern boundary, split into three nearly parallel ranges. The western range follows the Pacific coast, decreasing in altitude as it enters the Isthmus of Panama. The central range runs directly north until it terminates about one hundred miles from the Caribbean Sea. On its western side flows the Cauca, on its eastern the Magdalena, which unite at its termination and continue northward to the sea. The eastern range is more irregular and bears off to the northeast.

MAP OF THE
REPUBLIC OF
COLOMBIA

EXPLANATION.

CARIBBEAN SEA

PACIFIC OCEAN

A portion extends through Venezuela, whilst another portion continues as far as the Caribbean, where, near Santa Marta, it rises in snowy peaks 16,500 feet above the sea. From the eastern slope of this range countless rivers flow into the Amazon, the Negro, and the Orinoco. The Magdalena, which is practically the only highway in Colombia, has a dangerous bar at its mouth, but above this is navigable by steamers of light draught to Yeguas, a distance of some 630 miles. Here there is an interruption due to rapids, but above Honda small steamers continue the navigation to Neiva, and canoes are used even farther, making the total navigable length nearly 1,000 miles. From Honda to the sea the river falls between 800 and 1,000 feet, so is very swift, and were it not for its crookedness, the current would prevent navigation. Climates of all temperatures, from torrid heat to perpetual snow, are found in Colombia, and due to its broken surface it has two rainy and two dry seasons. For the Magdalena Valley, March, April, May, and September, October, and November are the rainy months, but the line between the seasons is not suddenly or sharply drawn.

Though there are a number of little fragments of railroads throughout Colombia, there is no railroad system proper, and where transportation cannot be had by water, dependence must be placed upon mules. Thus the capital of the Republic, a city of over 100,000 inhabitants is inaccessible by wheeled conveyance. There is said to be a poor wagon road from the river to the south of Honda, but it is seldom used.

CHAPTER III.

MONDAY, June 20, 1892. I looked out of the porthole of our little stateroom by daybreak this morning, and although I could see no land on account of a heavy mist, I knew that we were near the delta of the Magdalena. The sea was very muddy for many miles and covered with floating water plants and driftwood. In a short time the mist lifted and we began to catch little glimpses of the Colombian coast. We soon got our things together and came on deck, all excitement at the prospect of landing in a few hours. We finally came to anchor at half past eight about a mile from the land at Puerto Colombia. Savanilla was formerly the port, but the shifting sands have filled in the deep water there, so now the landing is several miles farther to the west. The harbor is a very exposed one, and I should think dangerous. There were several German and English steamers lying at anchor. We were shortly visited by the inspector of the port in a little cockle-shell of a tug with an excruciatingly shrill whistle, and about nine o'clock we got aboard of her and were taken ashore. On the tug were several passengers who had come from one of the other steamers, and on our way to the shore I made the acquaintance of one of them, a Mr. Lindauer of New York, engaged in business in Bogotá. Afterwards we saw a good deal of each other, and as he was familiar with the country, he was of great assistance to us on a number of occasions, and went to a great deal of trouble to help us.

We finally reached the landing, which was nothing but a few extremely slippery boards nailed to some worm-eaten piles in the

PUERTO COLOMBIA.
(After Millican.)

water's edge. Our satchels were tossed upon the landing, and we scrambled up as best we could, almost on our hands and knees. Once on top, we were surrounded by a perfect swarm of half-clad Indians and half-breeds of all sizes, who insisted upon carrying our things for us, whether we wished them to do so or not. Our trunks we could not take with us; we would have to get them at the custom-house in Barranquilla. The satchels of our fellow-travelers were inspected by the customs officers at the landing, whilst the rabble crowded around and examined everything critically. Upon showing my special passport, we were allowed to carry off our things without their being inspected.

Puerto Colombia is nothing but a collection of a half dozen wretched bamboo huts plastered with mud and thatched with reeds. The huts have no floors; there are stagnant pools of slimy water

in every direction, some even encroaching on the houses; a few
pigs wander listlessly about, and everything looks indescribably
filthy.   There is an iron screw-pile pier in process of construction,
alongside of which, when completed, it is intended for steamers to
lie, but it looked very weak to me.

We went ahead about one hundred yards to the railroad station,
where I got our tickets, and we boarded the train which was waiting. The road is a narrow gauge; the cars of two classes and somewhat of the appearance of our street cars. The freight cars are like
the little closed trucks used in transferring baggage across the New
York ferries.   Our train left for Barranquilla at half past nine, and
arrived there shortly after eleven. The distance is 18.5 miles. We
first followed the seashore for several miles, then turned to the right
and struck across country. The country that we passed through
was covered with a jungle of scrubby, thorny trees; no very large
ones, with now and then a small grove of cocoanut palms. In a
number of places rose large post-like cacti. The soil was sandy,
with a limestone outcropping at a few places. The Magdalena was
at its highest at this time; consequently the whole country was
flooded, and lakes and lagoons extended on both sides of the track.

As soon as the train moved off, I began to keep a sharp lookout
of the windows for birds. We saw large flocks of brown pelicans
(*Pelecanus fuscus*), numbers of white egrets (*Ardea egretta*), and
snowy herons (*A. candidissima*), small grayish herons similar to our
green heron but smaller (*Butorides cyanurus*), black vultures (*Catharista atrata*), flocks of large black ducks with a white spot in
each wing (*Cairina moschata*), pairs of large black and white stilts
with red legs (*Himantopus mexicanus*), great numbers of a species
of jaçana, dark, with a bright red frontal crest, and apparently all
the feathers in the last joint of their wings whitish (*Jaçana nigra*),
large crow-blackbirds, the females chocolate-colored (*Quiscalus assimilis*), long-tailed anis (*Crotophaga sulcirostris*), kingfishers,
larger than ours but with the same discordant rattle (*Ceryle torquata*), pigeons, ground doves, and quantities of flycatchers of dif-

ferent kinds. As we drew nearer Barranquilla I saw a flock of birds flying with rapid wing-beats, looking just like a flock of our doves; but as they veered off, the sunlight struck them and I saw that they were light green in color. They were parrakeets, the first birds that I had seen on the mainland answering my expectations as regards tropical birds. Later on, several flocks flew by the train near enough for me to hear their harsh, screeching notes.

Just after leaving the seashore, I noticed on both sides of the

HOTEL VICTORIA AND AMERICAN CONSULATE, BARRANQUILLA.

track among the trees a great many burrows with a little mound of earth thrown up around the entrance, and in each of these I could see a large blue crab (*Cardiosoma guanhumi*).

Our three guns, which were in their canvas covers and strapped in one bundle, had been passed by the inspector at Puerto Colombia, and we anticipated no more trouble about them; but, to our disgust, when we were leaving the station at Barranquilla, an old mulatto insisted on taking them to the custom-house. Arguments were of no avail; we had to give them up.

At the depot we took a carriage, a little open concern drawn by diminutive mules, and drove first to the Pensión Inglés, a hotel kept by a young Englishwoman, a Miss Hoare. Unfortunately for us, she had no vacant rooms, though she promised to let us have some on the following day. From here we drove to the Hotel Colombia, with no better success. Finally, at the Hotel Victoria we secured a couple of rooms. The hotel was a single-story building, one room deep, facing the street. Back of this was a large courtyard filled with beautiful flowers and fruit trees. This would have been a delightful place, had it not been for the fact that all the slops from the bedrooms were regularly thrown under the shrubbery. Back of this court and facing it was a row of bedrooms, and we were given two of these. The rooms were dirty, with cement floors, plastered walls, the under side of the roof for the ceiling. There was a heavy door in front, and one window in rear protected by wooden bars. It had blinds, but no glass. From its name we expected to find this an English hotel, but it was kept by a native woman, and practically managed by the negro waiter, Sam.

In our hurry in the morning we had left the ship without breakfasting, and here, according to the custom of the country, we did not get our breakfast until after twelve o'clock, so we were very hungry. Breakfast was served in the piazza facing the court. We had some strange dishes, none of them very good to my taste, but the coffee was excellent. The fresh meat is stringy and tough. Rice is well cooked, but is dark colored.

After breakfast I went out alone to attend to a few matters. I first called at the American consul's, but found him out. A few hours later I was told that there was a man in the house at the time suffering from yellow fever, which he had contracted at some mines up the river. This was rather pleasant for me, especially as I had entered the house. From here I went to the custom-house

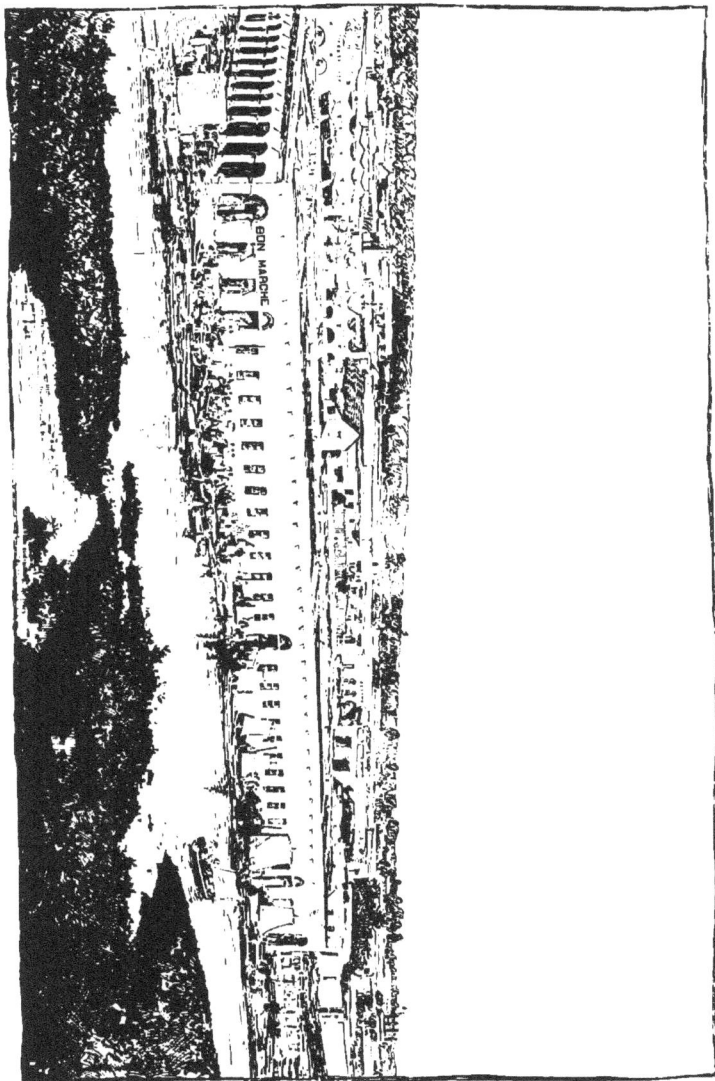

BARRANQUILLA FROM THE MARSH. MARKET IN FOREGROUND.

and after a great deal of wrangling succeeded in getting our guns.
The officials made but little objection to my taking the shot guns,
but haggled a great deal over the rifle. After my repeated assur-
ances that I had no warlike intentions, they finally gave it up to me.

I then went to the Banco Nacional and cashed a bill of exchange
for $500 in American gold, getting for it $1,000 in Colombian
paper currency. As a great part of this was given to me in small
notes, I had nearly a satchel full of money and felt very opulent.
The paper notes in circulation are the hundred, fifty, twenty, ten,
five, and one dollar or peso, and the fifty, twenty, and ten cents, or
centavos. The peso is regarded as divided into one hundred centavos, corresponding to our cent, and into ten reales, corresponding
to our dime. There are also three nickel coins, media, cuartilla,
etc., corresponding to 5, $2\frac{1}{2}$, and $1\frac{1}{4}$ centavos. Silver coins are very
scarce. Besides a few cuartillas I saw only two others, both fifty-cent
pieces, which I bought and kept as curiosities. Gold I did not see.
There are certain designations of currency which are apt to confuse
a stranger; for instance, there are terms which would nearly corre-
spond, if translated, to "hard" and "soft" dollars. A "peso
fuerte," or, as it is often called, a "fuerte," means a dollar of ten
reales, whilst a peso is generally taken to mean a soft dollar of eight
reales.

Later in the afternoon we drove around to the market and bought
some sleeping-mats, or "esteras." We had supper about six, and,
being tired out, went to bed early.

Barranquilla, although it covers a considerable area and contains
a population of over 20,000 inhabitants, does not amount to much
as a city except in a commercial sense. There are very few two-
story houses; nearly all are of one story, the majority built of bamboo and mud, plastered and whitewashed and thatched with rushes.
The floors are of mud or brick. All of the windows on the street
are protected by a framework of iron or wooden bars which pro-
jects about a foot from the wall. The houses are unprepossess-
ing from the outside, but as we passed along the streets we caught

glimpses through open doors of charming inner courts filled with beautiful flowers and plants. We noticed a peculiarity in the way that the furniture was arranged in the parlors. There were usually about six black rocking-chairs of bent wood in the room, and they were in the centre and facing each other in a double row, so close that they nearly touched.

The furniture of our bedrooms was meagre in the extreme; an enameled tin wash-basin and pitcher, a chair, an arrangement called a cot, but in reality a canvas stretcher fastened to a saw-horse. We

MARKET COURT, BARRANQUILLA.

spread our matting over this canvas, then a sheet over the matting, and the bed was made. Each cot had a good mosquito net suspended above it.

In the market we saw a number of curious things. The market building is a large one-story structure with an arcade on three sides

and a court in the centre. The side without the arcade is on
the water's edge, a side channel of the Magdalena. This front .
was crowded with canoes, all dug out of single logs, and some of
surprising size.

We saw a great variety of fruits. The sellers were mainly women,

DUG-OUTS ALONG THE MARKET FRONT.

who squatted with their wares exposed in front of them. The lower
classes here seem to be clean and good-looking; some of the women
are quite pretty. They wear dresses low necked and short sleeved
with very short waists, à la Madame Récamier; no head covering
beyond a shawl; their hair neatly arranged; a great many with
bright flowers in it. Children up to eight or nine go naked, or
nearly so. We saw several little babies, barely a month old, lying
on the sidewalk sleeping, naked and alone, with nothing under

them except perhaps an old piece of bagging or a few plantain leaves.

The water front of the market seemed to be the place of sale for fish. Although we saw no fresh fish, there were immense heaps of dried fish, split in the same way that our fishermen prepare mackerel. The greater part were small, but there were some large ones with immense scales. One that I examined closely looked to me exactly like the figures of the tarpon. It had the same general shape, the same thin, projecting under jaw, the large eye and scales, and the pointed projection from the dorsal fin. The Indian name was " sávalo," and they said that it came from higher up the river.

THE SÁVALO OR TARPON.
(From Goode's " American Fishes.")

Those that I saw were about two and a half feet long. Mr. Millican, in his "Adventures of an Orchid Hunter," p. 103, speaks of this fish, and says that he has seen specimens " seven feet long and two feet six inches in girth "! We also saw great piles of dried shrimps, which were sold by measure. They are eaten boiled with rice, but in my estimation the rice is sadly damaged by the addition.

There is in the town a street-car line, where little cars are drawn by sorry-looking mules, but it does not seem to be patronized. The streets are paved in but a few places; the rest is soft white sand, trying to the eyes when the sun is shining, and making all driving very heavy.

COFFEE SELLERS, BARRANQUILLA.

A great many small donkeys are used, and although they are not much larger than mastiffs, men ride them, sitting cross-legged like tailors, to prevent their feet from dragging. One man passed us perched on top of a little donkey, and with a live pig hanging on either side, squealing at every step.

There are barracks in the town, with a lot of dirty, unkempt soldiers who are continually tooting away on their bugles. Their call for taps is almost identical with ours.

There is in the town an electric light plant and also an artificial ice factory.

In the Hotel Colombia I saw a large scarlet, blue, and green macaw and a toucan with a serrate beak (*Pteroglossus sp.*). This bird assumed a most curious position when asleep, turning its tail up over its back and head instead of allowing it to hang as other birds do. In many of the houses along the streets we saw parrots, parrakeets, and troupials. Black vultures are abundant. They sit in groups in the cocoa palms, on the roofs and fences, and are continually flying down into the yards and streets to pick up refuse.

It was cloudy all day, and there were several showers. It was also hot, especially in the early part of the night.

Tuesday, June 21, 1892. We were awakened before daylight by the sound of music. It was the military band practicing, and although they selected such an unusual hour for their practice, I must admit that the music was excellent. Just about daybreak flocks of parrakeets began to fly over the town in a steady stream, and their incessant screeching put sleep out of the question.

We were up early, and after taking some coffee and bread, Cabell and I went down to the custom-house to get our trunks. Travelers' baggage up to two hundred pounds (as well as I remember) is admitted free of duty; anything beyond this must be paid for at an exorbitant rate. After waiting around for two hours, we got our trunks, and had them sent up to the Pensión Inglés, then went back to the Victoria, got together our things, and moved over. We had a good breakfast about half past eleven, and a little after two o'clock we took a carriage, and, Alice taking a book, and Cabell and myself our guns, we drove out a couple of miles into the country to have our first experience with South American birds.

We drove along a heavy, sandy road, with tracts of scrubby growth on either side, and here and there fields of a tall, thick, reedy grass. We saw no evidences of any crops. When we had gone out far enough, we turned out of the road, and left the carriage near an abandoned hut in an open field. We hunted around within a few hundred yards for about two hours, and saw great quantities of birds. I shot first, and killed a hawk that was perched in the top of a thick tree near the roadside. It saw me approaching, but was not shy, so I had no difficulty in getting within range. Before I shot at it, it uttered several times a shrill cry, and whilst doing so held back its head until its beak pointed vertically. It was about the size of our Cooper's hawk, its beak longer and not so hooked, its feet and claws weaker.   Its beak was light bluish, cere yellow, head and neck dirty white, a dark brown streak behind the eye, tail dusky with numerous narrow white bars, these bars becoming confluent at the rump, body and wings brown, below white with a buffy wash (*Milvago chimachima*).

Cabell then shot a curious kingfisher-like bird about the size of our catbird, but with a large head and heavy beak, which was slightly hooked at the tip, the hook being forked.   Around the base of its beak were stiff bristles pointing forwards.   Its toes were

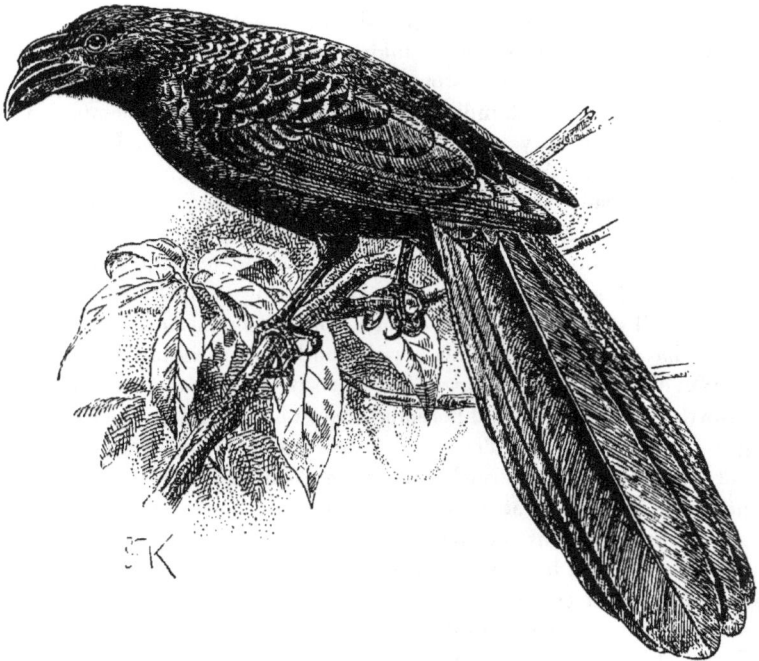

GROOVED-BILL ANI (CROTOPHAGA SULCIROSTRIS).

two in front, two behind; its tail-feathers narrow and weak. Its
head and upper parts were dusky, with buff edgings to the feathers;
there was a dark brown ring across the breast, with a whitish band
below; the throat was buff, with a rusty blotch in the centre. Beneath it was buffy, the flanks spotted with brown. There was a
white streak below the eye, and a white band at the back of the
neck. This was a rufous-throated puff-bird (*Bucco ruficollis*).

I then shot a crotophaga, probably smaller than our crow-black-
bird, but with a much longer tail, a curious high-arched bill; toes,
two in front and two behind; hackle-like feathers on its neck, and
of a uniform glossy blue-black (*Crotophaga sulcirostris?*). A little farther on Cabell shot a handsome flycatcher, much like our
great-crested, but larger, with a broad and large beak. Below it
was sulphur-yellow; above, rufous; its crown blackish, with a concealed light yellow patch, a white streak from its nostrils back
over the eye and entirely around the head. This was probably the
pitangua flycatcher (*Megarhynchus pitangua*). I saw several flocks
of parrakeets, — one of which lit near us, — and I started to creep
up on them; but they took alarm, and flew before I was within
range. They circled, and came back near Cabell; and he managed
to get one. It was about the size of a robin, but with a long,
pointed tail. Its beak and feet were light brown; its eyes, brownish
yellow. Its general coloration above was grass-green, with a trace
of blue in the primaries and secondaries; below, it was greenish
yellow; its upper breast, throat, and face were light greenish brown;
its forehead of a bluish gray (*Conurus æruginosus*).

I killed an oriole about the size of our Baltimore oriole; its beak,
wings, tail, and spot at the base of beak and under chin black; the
rest of its plumage a clear yellow (*Icterus xanthornus*); a small,
thick-beaked finch of a uniform glistening blue-black (*Volatinia
splendens*); a ground dove like those that we killed in Curaçao
(*Columbigallina passerina*); and a second one, somewhat larger,
and of a rufous color (*C. rufipennis*). Around a calabash-tree we
saw a couple of humming-birds, and Cabell managed to shoot one.

It was glittering green, almost exactly like those that we got in Curaçao, but its tail was forked (*Chlorostilbon angustipennis*).

We saw a number of partridges, and tried to get some, but failed. They were just about the size of our Virginia partridge; and, to my surprise, I several times heard them call "bob-white."

All of the birds that we killed were in poor plumage; they were evidently just beginning to moult.

We saw some brilliantly colored butterflies of various species.

I was surprised at the number of trees and bushes bearing thorns, — nearly all having thorns of different sizes. One tree, of large size and smooth, light green bark, had scattered over the trunk teat-like excrescences an inch or more in height and sharp-pointed, which would entirely prevent any one from climbing the tree. Some of the palms had very hard needle-like thorns, which would pierce the sole of a shoe; others had rows of short hooks arranged like the teeth of a saw.

As it was getting towards sundown, we turned back, and reached the hotel in time for supper.

In the evening the American consul, Mr. Neckius, and his assistant, Mr. Candor, called upon us. It was hot, and there were one or two light showers.

Wednesday, June 22, 1892. Cabell and I went out early to see about engaging passage and staterooms on the steamer Enrique, which was to start up the Magdalena on the following day for the head of navigation, Yegnas. After attending to this, we tried to get a trunk, so as to relieve the crowded condition of ours; but we could not find a suitable one, so finally bought a "pataca," a sort of bale covering, made of raw hide, the hair side out, which is used throughout the country as a case for transporting tobacco on mule-back. It is closed by being laced up with a raw-hide thong.

When we returned to the hotel, we repacked our things, leaving one trunk clear for skins.

After breakfast, we took our guns and started off on foot. We walked down the railroad several miles, and, taking it easy, came back at five o'clock. While walking along the track about a mile north of the town, we saw an alligator some four feet in length, which had been run over and killed by the train. About two miles down, we left the track, and turned into some scrubby, thorny woods to our left. Here Cabell shot a parrakeet of the same kind as the one that he had killed the day before, and a large pigeon, larger than our dove, but of the same uniform color throughout. The tips of its tail-feathers were whitish; but beyond this it had no distinctive markings. I did not skin this bird, as it was in poor plumage; and I did not get another specimen, so cannot identify it.

From here, we turned back to the right, recrossed the track, and went over to the river, where we found a great abundance of water-birds, the white-winged jaçanas, purple gallinules (*Ionornis martinica*), herons, ducks, etc. There was a skirt of small trees along the river, with here and there clumps of mangroves. Beyond the trees were reedy marshes extending out for perhaps half a mile. I had just reached the bank, and was walking along slowly, when a horrible-looking creature sprang up from under my feet and rushed off at a tremendous rate, stopping to look back at me when it had gone about thirty yards. I fired, and killed it. It was a lizard, over two feet in length, with very long and wide-spreading toes. It was brown, with darker markings on its sides, a conspicuous fin-like crest along its back and tail, and a light gray liberty-cap-looking growth at the back of its head. This was the basilisk (*Basiliscus americanus*). Later we saw quantities of them. They run with extraordinary rapidity, and stand higher from the ground when running than any lizard that I have seen. So rapid is the motion of their feet that they can actually run over the surface of water. This I saw repeatedly. I know of no other animal that can do this, except that I have seen frogs keep on the surface for a succession of rapid jumps; but frogs are web-footed, and these lizards are not. I saw several cross pools ten feet in width and keep on the surface for the whole distance. They also climb well. We saw them in the mangroves on branches overhanging the water.

As we passed under a low tree, one, frightened by us, sprang out on Cabell's back, and thence to the ground, giving him quite a start. We also saw numbers of other lizards, some striped green, blue, and yellow; other small ones, gray, with dark red heads.

Wading along the water's edge, we shot a pair of ibises, larger than our white ibis, but of the same general shape. They were of

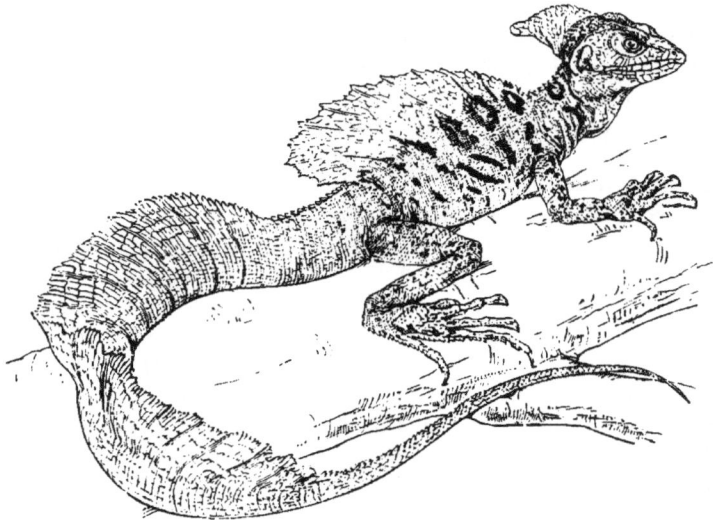

BASILISCUS AMERICANUS.

a dark glossy green, their legs, beaks, eyes, bare skin of face and gular space red (*Phimosus infuscatus*). We got several shots at flocks of parrakeets, and killed five or six, all of the same kind. They had been feeding on mangoes, and it was a difficult matter to prevent the soft yellow pulp that oozed from their beaks from soiling their feathers. We also shot some blackbirds of the same general shape as our red-winged ones, but smaller and with yellow

TURKEY-BUZZARD (CATHARTES AURA).
(From "Riverside Natural History," by permission of Houghton, Mifflin & Co.)

heads (*Xanthosomus icterocephalus*). I shot a small finch with a chestnut breast and a light gray back, but its plumage was in such a soiled condition that I did not preserve it (*Sporophila sp.?*). Flying about over the marsh we saw numbers of hawks, but we did not shoot any, as they would all have fallen into the water beyond our reach. They were large, dark brown with a conspicuous white rump, forked tail, and beak with a long hook (*Rosthramus sociabilis*). They quartered about like our marsh-hawk, close to the surface of the reeds. Among the rushes I saw some little birds conspicuously marked with black and white (*Fluvicola pica*). Their motions seemed to be just like those of our hooded flycatcher. They, too, kept out over the water, where we could not go, so we got no specimens. Cabell shot a hawk like the one that I had killed the day before, and I shot a second one very similar to the first, but with a brown head. It was probably a young one. " As wild as a hawk " is an expression of no meaning in Colombia; they are not at all shy, and it is an easy matter to approach within range.

We saw a few humming-birds, but got no shots at them. I also saw some turkey-buzzards (*Cathartes aura*), but they were scarce in comparison with the black vultures.

Birds were building at this season, and all were in bad plumage, so they were probably preparing for second broods. We saw ibises carrying sticks for their nests. On our way back we stopped at a little hut in a grove of cocoa palms, and I induced a small boy to climb one of the trees and get us some of the green nuts to quench our thirst with their milk. After throwing down some of them, he pulled out a nest from among the thick leaf-stems and threw it down to us. It contained two small spotted eggs nearly hatched, which were broken by the fall. The birds flew around uttering plaintive cries. They were the size of our scarlet tanager, and of a light bluish gray, darker on the wings and tail (*Tanagra cana*). The Indians called them " azulejo," which translates " bluebird " pretty closely.

After supper we skinned some of our birds, having a good deal of trouble with the parrakeets. It is difficult to get the skin of the neck to pass the head. It was hot all day, with a heavy rainstorm in the morning.

In the courtyard of our hotel there were several cages of parrakeets and troupials. One of the latter was a splendid songster, and imitated to perfection some of the bugle-calls. Whenever any one irritated it, it puffed out its throat until the hackle-like feathers stood out almost on end, and at the same time the pupils of its eyes contracted until they were mere points.

CHAPTER IV.

THURSDAY, June 23, 1892. We were busy packing in the early morning, as our boat was to leave at eleven, and at the last moment we were so hurried that we did not have time for breakfast, but snatched a few hasty mouthfuls and left. When we reached the Enrique, we regretted not having taken more time for our breakfast, for it was three o'clock when we finally moved off. It was very provoking to have to sit around and wait, but we could not help it, nor did any one seem to know for what we were waiting. Just as we were moving off we heard a great outcry, and, looking back, saw a passenger calling for us to come back for him; so we ran in to the shore, and he came aboard. Just imagine, in the United States, any one going at three o'clock to catch a steamer advertised to sail at eleven!

Whilst waiting at the wharf I noticed on shore great piles of what I thought were potatoes, but upon examination I found them to be vegetable-ivory nuts.

A great many kites, like those that we had seen the day before, flew about the steamer, and I saw them from time to time dip down gracefully and pick up some floating object from the water.

The Enrique, of which we give an illustration, was built by a Pittsburgh firm, and, like the Ohio River steamers, is a stern-wheeler, burning wood, of two to three feet draught, but high above water. On the lower deck forward are the boilers with wood stacked on either side; then comes the space for crew, freight, and live cattle for beef on the trip; then the engines. Forward, on the deck above, is piled the passengers' baggage, and this is where we spent the greater portion of our time when not driven in by the heat. Next come the staterooms, eight in number; then an open space, where we dined; and in rear the pantry and bathroom. Still higher is the pilot-house. The staterooms are small, perfectly plain, with a single canvas cot in each. No bedding is supplied by the boat, so a part of every passenger's baggage is a roll of matting, a pillow, and a mosquito net. The fare is sixty dollars in paper to Yeguas, staterooms ten dollars extra. The river steamers are compelled by law to carry a doctor. Ours was a native, and the captain was from Curaçao.

Our boat was in a side channel of the Magdalena, and had to go down about a mile before entering the main stream. This side channel was evidently the laundry for the town. The washerwomen waded out from its shores up to their waists, and pounded their soiled clothes on half-submerged drift logs which were scattered along. When we entered the main stream, we turned short about and headed due south. We went along slowly; the river was very high, muddy, and swift; and, besides, we had lashed to our side a large lighter, or "bongo," filled with extra freight that we had to take up the river with us. The country was inundated in all directions, and no high land was in sight. We saw thousands of water-birds of many kinds: white herons and egrets; large gray and black herons (*Ardea cocoi*), somewhat like our blue heron; a species of large tern, its body and tail appearing whitish, and its primaries, in strong contrast, black (*Phaëthusa magnirostris*). This tern we found abundant for four hundred miles up the river. The river was so high that no sand-bars were exposed, else we would have seen numbers of alligators; however, before dark we saw a few large ones on some logs. The native name is "caymán." I was told that there were several species. Shooting at them from the steamers was prohibited by law some years ago, owing to careless shooting by which a native woman on shore was killed; but our captain gave us permission to shoot when we got farther up the river.

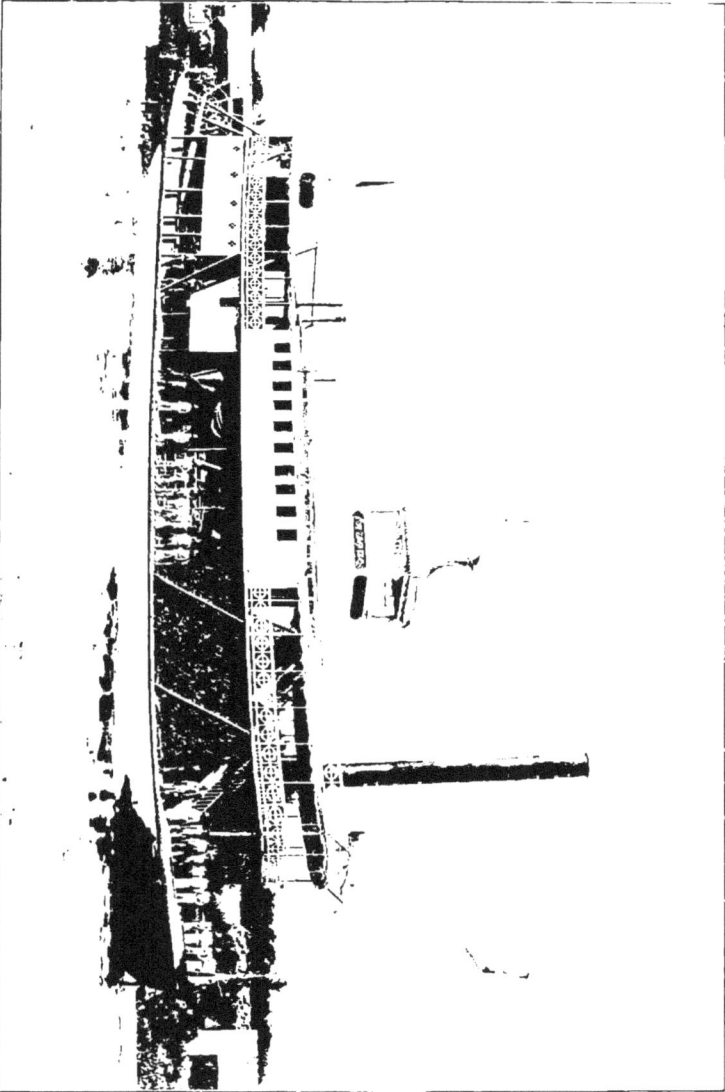

THE STEAMER ENRIQUE.

There is a good deal of ceremony at meal-times; no one takes a seat before the captain, and no one rises until he gives the signal. Should any one wish to rise before, he says, speaking to those present, "con su permiso," by your leave. The meals are served hurriedly by barefooted Indian boys, and were not so bad as we had been led to expect. There are but two meals a day, though

LAUNDRY AT BARRANQUILLA.

coffee is served soon after daybreak. The bill of fare is about the same for every meal, soup, beef and vegetables, "dulce" or sweets, which usually consists of some fruit such as green figs or "guayaba" skins, etc., boiled in syrup and served with coffee or chocolate and cheese. There was neither fresh butter nor milk. In every possible dish garlic is used and the majority of the dishes are colored yellow with arnatto. The vegetables are rice, potatoes,

yucca, plantains (boiled and
fried), and " ñames," or yams
as we would call them, though
they are entirely different from
the sweet potato to which we give
that name. The meat is always
in slices and is fried or stewed.
Roasts, joints, etc., are unknown.
The climate would not allow a
roast to be kept for even a few
hours. I witnessed one morning
the preparation of the meat for
the day. The cow was quickly
killed and skinned, then the flesh
was literally taken off in ribbons
until nothing but the bones were
left. These ribbons were wound
around slender rods, taken to the
upper deck, and exposed in the
sun. In a few hours they became
like pieces of sole leather. This
is called " tasajo " or jerked beef.
Before being cooked it is soaked
and beaten to soften it. The intestines, head, and bones of the
cow were turned over to the crew
of the bongo, who ate all with
relish, including the poor animal's
unborn calf.

Artificial ice is carried on the
up trip, but gives out about the
fourth day. Filtered river water
is used for drinking, and is fairly
good. The pilots are Indians,
usually old men, and are treated

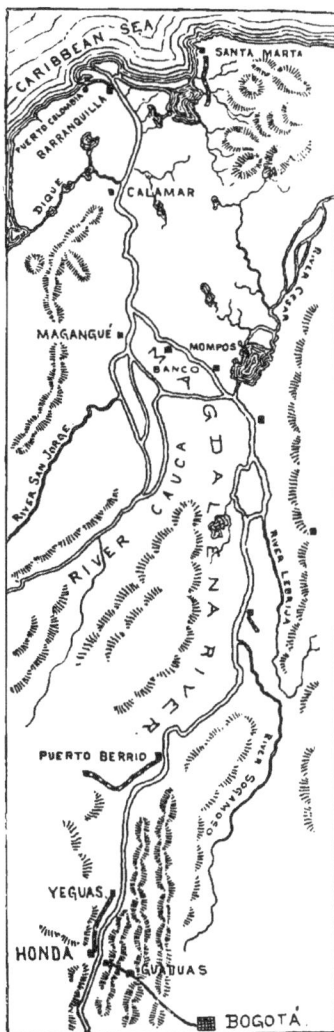

THE MAGDALENA VALLEY TO HONDA.

with great respect by the rest of the crew. There are no charts,
lighthouses, or buoys, and the water to the inexperienced eye looks
the same in all parts of the river, yet the channel is continually
changing and the pilots can tell at a glance when to cross from one

side to the other,
and when to keep
in the centre. The
boat stops three or
four times a day to
take on wood, which
is piled up along
the shore at conven-
ient places and sold
to the steamers by
the owners. There
are no wharves at
any place along the
river. The boat
simply runs up to
the shore, makes
fast to a convenient
tree, and puts out a
gang-plank. The
wood used for fuel
must be dry. It is
cut into lengths of
two feet, stacked in

COCOA PALMS ALONG THE MAGDALENA.

regular piles divided by upright stakes into small units called
"burros," which I suppose means a donkey-load. The price paid
is about fifty cents paper per burro. The wood is loaded by the
crew, who bring it on board on their shoulders, using a rope fastened around one wrist and held in the other hand to increase the
amount that they can embrace. They also usually wear a piece of
bagging over their head and shoulder as a protection against scor-

pions and insects that might be in the wood. This loading was a tedious process.

We also stopped a few times each day at little mud and thatch villages to take on or put off freight. The stops are of interminable length; no one seems in any hurry; after the freight is off or on they must have an hour's chat before starting, and when the signal sounds to start, the crew and passengers have gone off to

STOP AT BANCO.

make purchases or to trade, and must be waited for, so we really spend as much time in waiting as in traveling. We ran all night; but higher up the river, on account of snags and sand-bars, we had to tie up at night. It was fearfully hot, especially in the early part of the night, when it was almost unbearable in the little staterooms.

The majority of the passengers moved their cots out and slept on deck under heavy mosquito nets. Among the passengers we were pleased to find Mr. Lindauer and his cousin, on their way to Bogotá.

Friday, June 24, 1892. We were up by daybreak, and after having a cup of coffee went out on deck. At this hour the air felt cool and fresh, and it was by far the pleasantest portion of the day. The country through which we were passing was much the same as that of the preceding day; there were fewer cocoa palms and more mangoes and plantains along the shores. Magnificent unbroken forests stretched in all directions as far as the eye could reach. From time to time we passed little mud huts, thatched here with palms instead of rushes.

The quantities of herons and other waterfowl that we saw were incredible, the most abundant being the little snowy heron, which fairly swarms along certain portions of the river. Whilst in Barranquilla, I saw in one of the papers an advertisement of a New York dealer who offered to buy for cash the plumes of the snowy heron and of the white egret. It was accompanied by two wretched cuts of the birds with description of the manner of plucking and shipping the plumes. For those of the snowy heron he offered from $425 to $525 paper per pound, for those of the egret from $75 to $110 paper per pound. I was told that he had obtained somewhere near $10,000 worth of these plumes. As the snowy heron hardly ever has a dozen good plumes, and often only five or six, and as they have hardly any weight at all, one can easily imagine the numbers of birds that must have been sacrificed to the whim of fashion.

As we passed a marshy spot, we saw near the water's edge a herd of about a dozen reddish brown animals about the size of an average pig. They were capybaras (*Hydrochœrus capybara*), the largest of the rodent family. They paid no attention to our boat.

A little farther on, we saw walking about on a grassy spot a couple of large birds, looking much like our turkey, but having their heads covered with white feathers (*Chauna derbiana*).

Later in the day we saw a good many macaws, some green, blue,

and scarlet (*Ara aracanga*), others blue above and yellow beneath
(*A. ararauna*). This latter kind was the more abundant. They
fly heavily, like our crows, and usually by twos. Their long tails are
very conspicuous. Their harsh, discordant cries can be heard as far
as they can be seen, and were usually the first noises that we heard
in the early mornings. We saw quantities of wild ducks of several
kinds. Very often, when the flocks were near the forest, they flew
up into the trees when first alarmed. The largest kind, black with
white wing-spots, is called by the natives "pato real," royal duck,

CAPYBARA (HYDROCHŒRUS CAPYBARA).
(From "Riverside Natural History," by permission of Houghton, Mifflin & Co)

and is our muscovy (*Cairina moschata*). Another species, with
brown bodies and red beaks, stood in rows like soldiers along the
sand-bars (*Dendrocygna sp.*). I saw three kinds of kingfishers, all
in general appearance similar to our belted kingfisher. The largest,
which was larger than ours, was chestnut-red on the entire under

surface, including that of the wings (*Ceryle torquata*); the next in
size was marked like ours, but was glossy green instead of blue
(*C. amazona*); the third was a miniature of the second, about the
size of a large sparrow (*C. americana*). We saw all three kinds
enter and come out from holes in the river-banks. The first two
were very abundant, the third scarcer. We saw quantities of hawks
and large flocks of parrakeets, and I saw a single water-turkey or
snake-bird (*Anhinga anhinga*) flying high in the air. When the
crew were taking on wood at one place, they killed a couple of
slender snakes which were among the lower courses, but they were
thrown into the water before I could examine them.

In the afternoon, whilst we were stopping at a small village, a
native came up with a lot of fish in a dug-out canoe. They were
of two kinds : the first, a scale fish somewhat like a perch and of
about one pound in weight, he called "boca chica," little mouth ;
the other, a slender catfish, a "bagre," had the same smooth skin,
fleshy dorsal fin and beards that ours has, but its head was prolonged into a shovel shape almost like a duck's bill (*Platy-
stoma sp.*).

It was cloudy at times and hot, with a heavy storm at night.

Saturday, June 25, 1892. When I went out on deck this morning, I found that we were unloading freight at the town of Magangué. This is quite a busy little place, known for its annual
fairs. It lies in a strip along the river-bank with no high land
near. At this time many of the cross streets were flooded for a
portion of their length, and our boat lay alongside the sidewalk.
In a native canoe here I saw a skin very much like that of our otter.
The owner called it a "nútria," which is the Spanish for otter.

A short distance below Magangué the Magdalena separates into
two portions, inclosing a long island. Magangué is on the western
channel some leagues below the mouth of the Cauca. On the eastern channel is the town of Mompos, which was formerly of more
importance, but now, being inaccessible by steamers during the season of low water, it has lost a good deal. Upon leaving Magangué,

MAGANGUÉ FROM THE RIVER.

we returned to the forks of the river, where we picked up our bongo, which had been left there during the night, and then headed for Mompos.

About ten o'clock we stopped for an hour for wood, and Cabell and I took advantage of this to go ashore with a gun. Within fifty yards of the boat we found a small tree covered with fringy-looking flowers, and around these some humming-birds were feeding. In a few minutes we killed six, two of one kind and four of another. The first were of moderate size, bills broad at the base, reddish with dark tips. They were green above, throats metallic green, under parts ashy, tail, including the upper and under coverts, rufous, the retrices with narrow bronze edgings (*Amazilia fusicaudata*). The second kind were green above, throats glittering green, lower part of breast grayish, a white patch on the belly, under tail coverts

green with gray edges, tail forked and blue-black, the two central feathers greenish (*Cyanophaia goudoti*).

AMAZILIA FUSCICAUDATA.
(From Elliot.)

Near here, Cabell shot
into a flock of parrakeets
in a mango-tree, and killed
three. They were different
from the others that we had
gotten, being smaller and
of a brighter green, the
alula principally blue, under wing-coverts light yellow, upper coverts brown-
ish green, an orange chin-spot, bill and feet flesh-color (*Brotogerys jugularis*). There was a peculiar point on the inner web of
the third primary. These little birds hang head downwards on the
mangoes, and tear at the soft yellow pulp until nothing but the
seed is left. When a flock is in a thick foliaged tree, although
they may be very noisy, they are sometimes difficult to see, as their
colors harmonize closely with those of the leaves. Cabell also shot
an "azulejo" (*Tanagra cana*), a male in fair plumage. We
caught here some beautiful butterflies,
some morphos especially, large ones,
brown beneath with round eye-like
spots, and above rich azure. Others
with swallow-tails were striped metallic
green and black, and others scarlet
and black.

CYANOPHAIA GOUDOTI.
(From Elliot.)

Throughout my stay in Colombia I
had untold trouble in keeping butterflies. There was a minute red ant on
the boat which soon found anything to
eat, and destroyed it in a few minutes. Some butterflies that I had
put in a tin box the day before were nothing but fragments when
I examined them. The only sure way is to put the box on a little

pedestal in a basin of water, and to examine it every few hours to see that the water has not evaporated. After breakfast, we prepared our birds as the boat went along, shot at alligators from time to time, and tried fishing when the boat stopped, but got no bites. We saw birds in great abundance, and, among new ones, some large green parrots. They, like macaws, fly in pairs; but their manner of flight is as different as possible. They have a rapid, tremulous wing-beat, exactly like that of our leather-wing bat. Speaking of bats, there are a great many along the river, and at nightfall we saw them flying about close to the surface of the water. Some are much larger than ours, with longer and more pointed wings.

Late in the afternoon we reached Mompos, and shortly after had the chagrin of seeing the mail-steamer, which left Barranquilla the day after we left, pass us on her way up. Mompos is an old town, with some ruins of an ancient cathedral. We bought here from Indian women who came on board some dulces, guava jelly, limes preserved in syrup, etc. An Indian offered to sell me for fifty cents a half-fledged blue and yellow macaw; but whilst I was thinking it over the bird uttered one of its horrible squawks, which decided me to do without it. It was apparently full-sized, and had a few blue feathers above, but below was naked. I saw in Mompos a leper, the first I had seen, although I had heard that there were many in the country. At a number of places along the river we saw a form of skin disease which was called "carate." In some cases the dark skins of the Indians were covered with light spots and blotches; in others the spots were bluish black. The hands were more affected than other portions of the body. There was nothing malignant about this, simply a discoloration of the skin similar to scars left by scalding, without any contraction.

We ran all night. It was hot during the day, and hotter at night.

Sunday, June 26, 1892. We woke this morning early, at a place called Banco. It is a small village, with the usual cathedral, situated on a hill or bluff of red clay. There was a crowd of natives at the landing, with sleeping-mats and other articles for sale. Here I purchased for forty cents a large and prettily marked tiger-cat's skin. Later in the day we stopped several times for wood, and at one place we went ashore. We saw many wren-like birds, some resembling our Carolina wren, but as large as a cat-bird. Cabell shot a second "azulejo." During the day we had a

CATHEDRAL AT BANCO.

great many shots at alligators, but struck only a few. Among new birds I saw several small flocks of roseate spoonbills (*Ajaja ajaja*), and some immense flocks of wood ibises (*Tantalus loculator*). It was clear and hot during the day, but cooled off a little at night, so that we could go to sleep without the preliminary Turkish bath. Cabell saw to-day, floating in the river, a dead snake about ten feet in length.

COLOMBIAN SCREAMER (CHAUNA DERBIANA).

Monday, June 27, 1892. Upon waking early I found that we were unloading at a little group of huts, and as I heard a great many birds, I hastily dressed and hurried ashore with my gun. Within a few yards of the boat I shot one of the medium-sized kingfishers, a male, marked like ours with a chestnut belt, but glossy green above (*Ceryle amazona*). I saw here a flock of little

short-tailed parrakeets, as small as sparrows (*Psittacula conspicillata*), and some little swallows about the size of our bank-swallow,
with white bodies and dark wings (*Tachycineta albiventris?*). I
had to hurry back to the boat before I could shoot any more, and
on our way up the river I skinned the kingfisher. Later in the
day we stopped again, and I went ashore, but found it so intensely
hot that I soon came back. I saw here, with some chickens, a pair
of the turkey-like birds that I had seen on the 24th. They had
red legs, with long straight toes and claws, and spurs on the last
joint of their wings. Their general plumage was black; their
faces white, with a red ring around the eyes, and a feathery horn
on each side of the head (*Chauna derbiana*). In the afternoon
the boat stopped for wood, and we went ashore again. This time I
got a fine pigeon, a male, as large as our domestic pigeon. It had
a bluish rump, forehead, and throat, purplish back and wings, a
metallic green nape, red feet, eyes, and lids (*Columba rufina*). I
saw during the day several caracara eagles (*Polyborus cheriway*),
and with my glass I could plainly see the brilliantly colored skin of
their faces. All day long we saw enormous flocks of ducks, wood
ibises, and parrakeets, and quantities of white herons, white egrets,
cocoi herons, blue and yellow macaws, parrots, hawks, kingfishers,
and a few fish-hawks (*Pandion haliætus carolinensis*). We fired
many times at alligators, and saw some very large ones. We tied
up to the shore at night, as the river had become too full of snags
and bars to navigate except by daylight. We struck sand-bars
twice in the afternoon, but fortunately got off easily. For the last
two days we have had lovely views of blue mountains. To-day
they were to the west of us. It was clear and very hot during the
day; but we had a shower at bedtime.

Tuesday, June 28, 1892. We were up early, and at the first
stop for wood went ashore with our guns. We found the land to
be only a few inches above the level of the river, of a soft black
mud, and near the water covered with a heavy growth of large
canna-like plants, with red and yellow flowers. Around these were

feeding some humming-birds, and Cabell shot a pair. They were larger than any that we had met before, and had long curved bills, the lower mandible yellow, the upper dark with a yellow streak on each side. Above they were metallic green, the upper tail-coverts with light buff edgings, the throat rufous, under parts buffy, central tail-feathers green with whitish tips, the others rufous with whitish tips and a blackish subterminal bar. There was a light buff streak from the gape and another from behind the eye (*Glaucis hirsuta*). One of these, a female, had a number of white feathers scattered among the green ones of the back. I shot here one of

GLAUCIS HIRSUTA.
(From Elliot.)

the rufous-tailed humming-birds (*Amazilia fuscicaudata*). From this place we pushed on about fifty yards, until we reached the edge of the forest, and here we found birds in abundance. Cabell shot first and killed a large bird nearly the size of our crow. This was a male. It had an oriole bill, black with a coral red tip, a light blue excrescence on each side at the base of the lower mandible, a flesh-colored excrescence on its forehead, and light blue skin around and back of its eye. Its feet were crow-like and black. Its under parts, head, neck, and wings were black, the feathers of the neck with white bases. From its forehead sprung three long filamentous feathers. Its upper wing-coverts, scapulars, centre of its rump, and under tail-coverts were rich chocolate. Its tail was clear yellow with the exception of the two central feathers, which were black, and which in this specimen extended only halfway down to the tip of the tail (*Gymnostinops guatimozinus*). The natives called it an " oro péndola," gold hang-nest; but they apply this name indiscriminately to all the oriole family that build pendent nests. About the same time I shot another, very similar in style and pattern of coloration, but of about half the size. This was a female, its bill plain ivory without excrescences, and the feathers on its crown only slightly prolonged, otherwise its coloration was the same (*Ostinops decumanus*). The two kinds were together in a large straggling flock. Still later I shot a third, smaller yet, black with a black tail, a clear yellow rump, under tail-coverts, and wing-spot. Its bill, which was slightly curved, was a pinkish ivory, and the feathers of the crown were slightly prolonged (*Cassicus flavicrissus*). This also, like the first two, had white bases to the feathers of the nape. These birds build together in communities. A number of times, along the river, we saw in large detached trees a dozen or more of their nests hanging like stockings from the extremities of the branches. As I shot the second, I heard the

"ORO PÉNDOLA" (GYMNOSTINOPS GUATIMOZINUS).

PSITTACULA CONSPICILLATA. *Lath.*
Blue-rumped Parrakeet.

Mintern Bros. Chromo lith. London

harsh screams of some macaws ahead of me, so I pushed on through the trees, and got a long shot at one which fell screaming in a thorny jungle. I forced my way into it, and as I picked it up it bit my thumb until the blood streamed, and before I could choke it off I began to be afraid that my thumb would be cut in two. Its cries attracted its mate, which I also shot. They were smaller than any macaws that I had seen, and were in wretched plumage. Their general color was a grass-green, bluish about the head, a reddish brown stripe on the forehead, primaries blue above, reddish beneath, under wing-coverts scarlet, tail reddish at base, then green, then blue, but reddish beneath, skin of face white with lines of bristly black feathers, beak black, feet dark (*Ara severa*).

A little later I shot a pair of the small parrakeets that I had seen for several days past. They were miniature parrots, no larger than sparrows, a bright grass-green, with secondaries, upper and under wing-coverts, rump, and a ring around the eye a deep blue, beak and feet flesh-color (*Psittacula conspicillata*).

I also shot a tanager, which the natives called a "cardinal." It was like our scarlet tanager in size and distribution of color, except that the scarlet, which was beautifully clear on the rump, grew darker towards the head until it became a dark garnet. The plumage was velvety, especially the black of the wings. The upper mandible was black, the lower a light horn-color (*Ramphocelus dimidiatus*). Cabell then shot a small puff-bird about the size of our pewee, but with a larger head and weaker tail. Its upper mandible was forked at the tip like that of the one that we shot at Barranquilla. It was black above, white below, with a black collar, white specks on the forehead, a white spot on the scapulars and a little white on the rump (*Bucco subtectus*). Just as we were getting on the boat, he shot a beautiful little bird about the size of our chipping-sparrow, glossy blue-black above, with a yellow forehead and bright yellow below (*Euphonia crassirostris*). This was a male, and in better plumage than any bird that we had gotten so far. The female, as I found out later, is of a plain greenish yellow.

After the boat started, I was busy for several hours skinning the
birds. The macaws were especially troublesome, as the skin of the
neck refused to pass over the skull.

In the afternoon the boat stopped again and we went ashore, but
it was so boiling hot that very few birds were stirring. Cabell,
who was some distance ahead of me, fired, and as I came up he
called out that he had killed a humming-bird as large as a tanager.
It was certainly a beautiful bird, and its metallic plumage and long
bill gave it a slight resemblance to a humming-bird. It was a jacamar, brilliant metallic green and bronze above, including the two
central tail-feathers. The remaining tail-feathers and the under
parts were rufous. Its throat was white and was separated from the
breast by a band of the same color as the back (*Galbula ruficauda*).
I saw here a pair of toucans, and got a shot at one, but failed to get
it or to see whether I had hit it or not. Its breast was dark red;
its other colors I could not distinguish. I also saw in the forest a
number of dark reddish squirrels with white bellies. They were the
size of our gray squirrel and were extremely gentle, allowing me
almost to touch them with my gun-barrel as they sat watching me.
On my way back to the boat a bird fluttered up from the thick
grass in front of me, and I got it by a snap shot, but my heavy
choke-bore unfortunately spoiled it as a specimen. It was a species
of whippoorwill, just about the size of ours, and, like ours, had
bristles along its gape. It had a white throat-patch, and beneath
was marked just like our night-hawk, but the ground color was
more reddish brown. Its wings and tail were somewhat like a
whippoorwill's, the wings with a light buffy spot on the primaries.
Its back was mottled and the scapulars had buffy outer edges
(*Nyctidromus albicollis*). Several times at night along the river
I heard the cry "whip-poor-will," and others very similar, but I
do not know what bird uttered them.

At this place the steward of the boat came up to me with two
dirty white eggs just the size and shape of those of our yellow-
billed cuckoo. Showing them to me, he said, "azul, azul" (blue,

blue), and going off he returned with a saucer of wood ashes and a moist rag, and began to rub the eggs. In a short while all of the white disappeared and they became the color of a robin's egg. He said that they were the eggs of the ani.

In the afternoon I skinned the birds, and we shot a good many times at alligators. The river was now very crooked and swift and full of sand-bars and snags, so at dusk we tied up for the night. At this place we saw two long-tailed monkeys make off through the treetops as we came up. We saw quantities of birds all day, blue and yellow macaws, ducks, herons, ibises, parrakeets, spoonbills, etc.

I was fighting red ants throughout the day. The few butterflies that I had captured, I tried in every way to save. They were put in tin boxes with camphor, but whenever they were left for two hours I invariably found them literally swarming with ants, their heads and bodies eaten off, and their wings coming to pieces. Nothing but putting them on a tumbler in a basin of water protected them. This was impracticable for bird-skins, and I was afraid that I would lose them all. I put the skins in the tray of my trunk, which I suspended by strings from the ceiling, but by night I discovered the ants traveling up and down the strings in an unbroken column. After this I rubbed the strings with kerosene oil and carbolic acid, and tied lumps of camphor to them, but the ants were not delayed in the slightest. I finally borrowed from the steward three soup-plates, which I filled with water and placed in the centre of each a tumbler; on these three pedestals I put my tray, and the ants were baffled at last.

It was clear and very hot, especially in the early night, but we were not troubled by mosquitoes.

Wednesday, June 29, 1892. Cabell was taken with a slight fever last night, caused by going out in the hot sun yesterday afternoon. He felt badly all day, so did not leave the boat. At our first stop, Lindauer and myself went ashore and killed a number of birds. I shot first a pair of the little blue-rumped parrakeets (*Psittacula conspicillata*), a male and female. The female is plain grass-

green without any blue. In a marshy spot near a little stream, I
shot one of the black and white birds that I had seen in the marshes
at Barranquilla. It was a male, a little smaller than a pewee, white,
with wings, tail, back of head, and centre of back black (*Fluvicola pica*). I also shot three more jacamars (*G. ruficauda*) and a
puff-bird like the one we got at Barranquilla (*B. ruficollis*). Lindauer shot a couple of flycatchers; the first, a male, smaller than
our bee-martin, yellow below, brownish olive above, crown brown,
with a large yellow and orange patch, white streak from nostrils
above eye to back of head, and throat white (*Myiozetetes cayennensis*); the second, a female, about the size of our great-crested
flycatcher, plumbeous above, a small orange crown-patch, throat and
breast grayish, and below light yellow (*Tyrannus melancholicus*).
He also shot a most peculiar and beautiful little bird, a male in fine
plumage. It was about the size of a wren, but with an extremely
short and awkward-looking tail. Its legs were white with a scarlet
ring above the tarsus, its head rich golden yellow becoming orange
with traces of scarlet at the back. The rest of its plumage was
glossy blue-black. Its eyes were white with fine red lids, and
its bill light yellow (*Pipra auricapilla*). At this place I saw a
flock of certainly five hundred of the orange-chinned parrakeets
(*Brotogerys jugularis*) in a mango-tree near the boat.

After leaving this place, we stopped no more until we tied up for
the night; so I spent the rest of the day in skinning the birds and
shooting at alligators. Every sand-bar, or "playa" as they are
called, was sure to have a number on it. They generally lie in the
sun with their mouths wide open, the upper jaw making an angle
of forty-five degrees with the lower. When shot at, they sometimes
slid off into the water like terrapins from a log; but when they
were well up on the playa, they rose deliberately to their feet and
walked off, their bodies looking as high from the ground as that of
a dog.

All day long the river was very crooked; there were bluffs of
red clay along the shores; the country was not so marshy, and we

LOOKING DOWN THE MAGDALENA FROM BANCO.

saw no ducks or white egrets, but numbers of macaws, parrots, kingfishers, and wood ibises. The doctor gave Cabell a sudorific, and at night he was much better. We spent a very hot night, tormented by mosquitoes.

Thursday, June 30, 1892. We made an early start this morning, and did not stop until we reached Puerto Berrío, about ten o'clock. This is a village on the western bank of the Magdalena, and is the starting-point for the Antióquia Railroad, which is destined to reach Medillin, the capital of the department of Antióquia, but which now terminates at Pavas, about twenty-five miles from the river. Here Lindauer and myself went ashore with the guns. Cabell, although feeling well, thought it best to keep out of the sun. We went back a short distance along the railroad track; but it was rather late in the day for the birds to be stirring, so we saw only a few. I got two new ones: the first a tanager, a male just the size of the "cardinal." It was velvety black, with a beautifully clear yellow rump, its bill light horn-color with darker cutting edges (*Ramphocelus icteronotus*). The second was a humming-bird, a female, green above, the rump and tail-feathers bronzy, the lateral tail-feathers growing darker towards the ends and tipped with white. Below it was gray-ish, with a few metallic green and blue feathers on the throat (*Polyerata ama-bilis*). The natives call humming-birds "chupa flores," flower-suckers, and some-times "pica flores." Several hours later in the day the boat stopped again and we went ashore, but it was too scorching hot for anything to be stirring. I shot a large oriole, about the size of our robin, with a black beak, face, chin, and wings, and black and yellow tail, the rest of the plumage yellow. It was in such poor plumage that I did not preserve it, so now have to regret not being able to identify it.

POLYERATA AMABILIS.
(From Elliot.)

On my way back to the boat I saw up a small tree what I thought was a very large snake, but upon closer examination I found it to be an iguana, which I shot and carried back with me. It was forty-three inches in length, the greater part of this being taken up by its tail, which tapered to a point and was striped with broad bands of gray and black. Its body, which was about the size of our rabbit's, was green with black marks. Along its back was a row of leath-ery spines (longer than in the species figured), and beneath its throat was a pouch or dewlap. I skinned its body, and got one of the bongo men to cure the skin for me by rubbing it with wood-ashes. Its flesh, which is eaten by the natives, looked good, and I noticed that it had the same odor as that of our bull-frog. At this place there were a few Indian huts, and around them a small grove of cacao-trees, from which chocolate is made. They were not over twenty-five feet high, smooth barked and big leaved. The fruit looked very much like an oblong warty squash, and grew close to the main trunk and large limbs. They were about eight inches long, some green, others a deep purplish red, and when cut open showed a white pith in which were imbedded bean-like seeds the size of our lima beans but thicker. These, when ripe, are taken

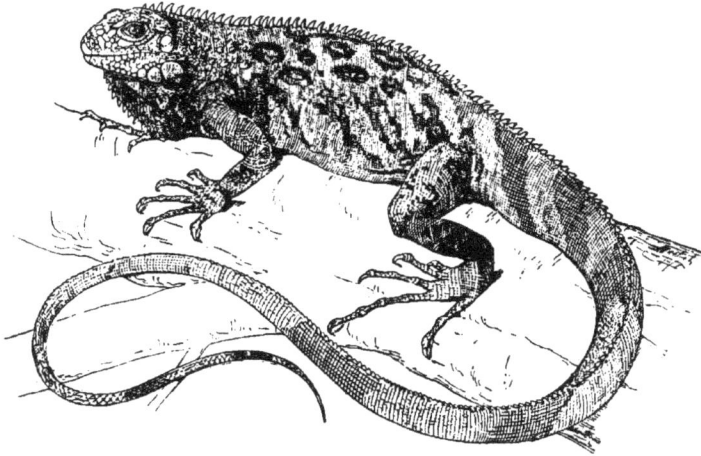

IGUANA TUBERCULATA.

out, roasted, and then ground between two stones, mixed with coarse sugar, and the result is chocolate. Hung up against one of the huts to dry, I saw several peccary skins of the plain unbanded species (*Dicotyles labiatus*). I was told that they were common in the forest here.

Lower down along the river the native huts are made of a wattle of split bamboo, or small sticks, daubed with mud and thatched with palm-leaves (see page 55), but here the walls are made in a different manner. The large cane or bamboo, the "guaduas," which often is six inches in diameter, is taken and partially split in a number of places about an inch apart, after which the whole tube can be opened out, making a very rough plank from a foot to eighteen inches in width. These are lashed to the framework with bark or slender vines. We also saw many huts with nothing but a roof and the four corner posts, protection from the sun and rain being all that was required.

The natives along the river are, as a rule, cleanly, amiable, inoffensive, and very indolent. All carry the "machete," a long and heavy sword-like knife, which is the universal tool. It is about thirty inches long, sharp on one edge, the back being very thick, and the blade widens from the handle until near the point, where it is sometimes five inches broad, then tapers suddenly. It is used like a cleaver. Those that I saw were made in England and in the United States. They are sometimes carried in a heavy leather scabbard, sometimes in a small loop of leather tied around the waist. I saw a few axes, but they were all of the old Spanish pattern, like those shown in the old illustrations of "Robinson Crusoe," the blade fan-shaped, with a ring at the back for the insertion of the handle.

Notwithstanding the tremendous forests, lumber of all kinds is scarce and dear. There are few, if any, saw-mills; boards are usually sawn out by hand, and a plank ten feet long, a foot wide, and an inch thick sells for a dollar in gold.

Although the natives are indolent, they can work, for the bongo men sometimes toil day after day under the broiling sun for a month or six weeks, poling their heavy bongos up the Magdalena. And, after all, a living comes so easily to them, their wants are so few and so easily supplied, that there is no incentive for them to work. When a native wishes to set up a house for himself, he selects a convenient spot along the river's bank, then with his machete cuts down the bushes and vines and girdles the larger trees over an acre or two, clears off the débris by fire, then plants a hundred plantain shoots. In a little over six months the plants will have fruit ready for food. One bunch, which can be bought along the river for a real, will keep a man in food for ten days. The plantains are eaten green or ripe, boiled, baked, fried, or raw, and are a fair substitute for potatoes and bread. As soon as the bunch of fruit is cut off, the plant is cut down close to the ground, and it immediately puts up fresh shoots which bear again in six months, and so on. The natives call plantains "plátanos," and bananas they call "platanitos," little plantains. The bananas that we got in Colombia were

A BONGO OR CHAMPAN ON THE MAGDALENA.
(By permission of Bureau of American Republics.)

among the most delicious of fruits. They were small, with a skin as thin as a kid glove, and of an exquisitely delicate flavor, incomparably superior to those that we have. These will not bear transportation. From seeing the bunches before our fruit stores, I had always thought that bananas grew pendent on the bunch, but they grow with their free ends pointing up. The natives raise a little corn, but there is no systematic method of planting or cultivating it. The difference in cultivation is shown by the ears, on which the grains are irregularly distributed, and not in long parallel rows as in our corn. As there are no mills, they grind the little corn that they need between two stones, the same two stones that are used in making chocolate and also in grinding coffee. The river supplies them with fish and turtle in abundance, and they easily

CITRON-BREASTED TOUCAN.

trap different birds near their huts. They need but few clothes, they raise enough tobacco for their own use, and the native rum, "aguardiente," costs about the same as our cider. Their household furniture is limited to a few hammocks, two or three earthenware pots, and a supply of calabashes and turtle shells which serve as dishes and spoons.

In the afternoon we came to a portion of the river called "Angostura," or narrows, very narrow and swift, where even with a full head of steam we barely crept along. Here I saw a great many turtles and alligators, large flocks of macaws, and some roseate spoonbills. Late in the afternoon we stopped for wood and I went ashore, but did not take my gun. Lindauer took one of the guns, and in a few minutes returned with two new birds. The first was a very fine toucan, a female in good plumage. It was about the size of our crow, had a very large, finely serrate beak which was brilliantly colored with black, white, green, blue, and yellow. Its eye and the skin of its face were a beautiful peacock-blue, its feet light blue. Its general color was black, breast, throat, and face light yellow, becoming white on the cheeks, and separated from the black of the under parts by a bright red belt. Its tail was black and square, the upper coverts yellow, the lower bright red (*Ramphastos citreolæmus*. (See frontispiece.) The second was a parrot, the size of a small pigeon, a female in poor plumage. Its beak was black with a coral-red spot on each side, general plumage green, and head and neck blue, ear-coverts black, a few rosy feathers among the blue of lower throat, the four central tail-feathers green with blue tips, the others blue, rosy at the base. The under coverts were pink with blue stems and yellow tips, the edge of the wing pink and yellow (*Pionus menstruus*). I found both the toucan and parrot difficult to skin on account of the smallness of the neck. The colors of the beak and skin of the toucan faded in a few hours. The nostrils of the toucan were not in the beak proper, but in the crease between the base of the beak and the frontal feathers. The "pope's nose" of the toucan was longer than that of any bird that I have skinned, and it is so freely jointed that the bird can move its tail in any position. It is owing to this structure that when roosting the toucan can turn its tail over to cover its back and head.

The boatmen killed in the woodpile here a scorpion, plain olive-green in color, and the size of a small fiddler crab. We tied up for the night. It was hot, but we were not troubled by mosquitoes.

Friday, July 1, 1892. We made an early start and did not stop until late in the forenoon, when it was too hot to find many birds. I went ashore and killed a curved-billed humming-bird like

COLLARED ARAÇARI (PTEROGLOSSUS TORQUATUS).

those that we had shot on June 28 (*Glaucis hirsuta*), and a pair of new toucans, smaller than the one that Lindauer killed. Their tails were longer and the feathers graduate like those of our cuckoo. Their beaks were deeply serrate, the upper mandible yellowish white with a black tip, a black streak on top, and a reddish mark at the side of base; the lower mandible black, and both bordered at the base by a white line. The skin of the face was scarlet, the eyes yellow, and the feet olive-green. The head and throat were blue-black, a brown collar at the back of neck; back, wings, and tail greenish black, rump scarlet, below yellow, orange on the breast, a black spot in centre of the breast, and lower a black and red belt, the thighs brown (*Pteroglossus torquatus*). Both were females in poor plumage. Their tongues were bristly, like a worn-out feather. The remainder of the day I did but little.

The river-banks became higher and gravelly, the water much colder, and fewer alligators were seen. We dropped our bongo, so made better time, and taking advantage of the moonlight, we ran until nine o'clock, and finally tied up about fifteen miles below Yeguas, our destination. It was very hot all day, but cooled a little after sunset.

CHAPTER V.

SATURDAY, July 2, 1892. We made an early start, but stopped for wood a few miles below Yeguas. I was busy getting our baggage together, but went ashore at this place. I saw no birds, but found scattered about over the ground a number of land shells,

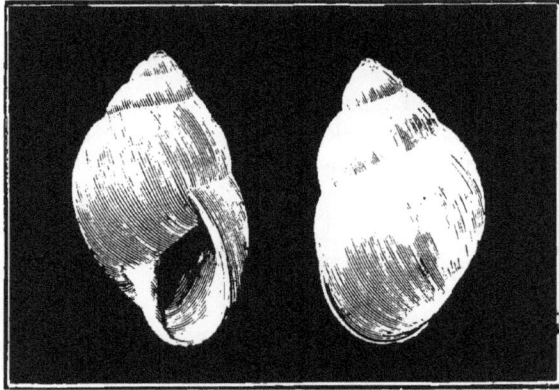

LAND SHELL FROM NEAR YEGUAS.

white, with rosy lips, the largest that I had ever seen, larger than lemons, some being four inches long (*Bulimus oblongus*, Müll.). I brought back several with me. I was told that the animal inhabiting these shells lays an egg much similar in size, shape, and color to the eggs of the little ground dove. Shortly after I came on

THE DIAMOND RATTLER.
(From "Riverside Natural History," by permission of Houghton, Mifflin & Co.)

board, some of the men came down to the boat, dragging a very large rattlesnake, which they had just killed near the spot where I had picked up the shells. It was not so brightly colored as those that we have in Virginia, but was rusty brown, with a series of dull yellowish, diamond-shaped marks along its back. The native name for rattlesnake is "cascabel."

Just before reaching Yeguas the river becomes very rapid, and curves to the left for almost half a circle. Yeguas, which is on the western bank, is a collection of four or five bamboo and thatch huts upon the top of a gravelly bank, some twenty feet above the water. One of these huts serves as a station for the Dorada Railroad, which runs from here to Honda, about fourteen miles above. We arrived at ten o'clock, just half an hour too late for the morning train, so were compelled to wait on board until half past three. The road is narrow gauge, the cars small and not very clean, and the country hot and dusty. At Yeguas the character of the country changes abruptly, the heavy forests disappear; their place is taken by level plains, good examples of geological terraces, with here and there high, flat-topped, and barren hills. The strata in the hills lie horizontally, and erosion has produced the same style of landscape as seen in many pictures of Arizona. Upon leaving

Yeguas, the train first goes up a steep incline, until it gets upon the level terrace, where it runs for some time at a fair rate of speed. This plain is in parts several miles broad, covered with a very rank sort of grass or broom straw; and scattered here and there are clumps of palms. A great many cattle were feeding about. Along here on the telegraph poles I saw a number of small hawks, apparently the same as our sparrow-hawk, and some large buzzards, larger, perhaps, than our red-tailed hawk, with dark reddish brown wings (*Heterospizias meridionalis?*).

After going about five miles, we heard a great whistling and

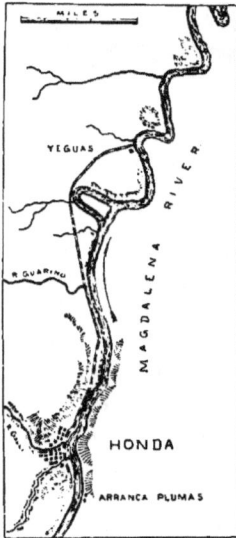

tooting of the engine, and looking out saw that we had just run over a cow. Instead of stopping the train, the engineer tried to pull it over the cow; so, after she had been dragged several hundred yards, and had rolled from one car to another, until she reached the centre of the train, the rear wheels of a truck were thrown from the track, and we had to stop. By the help of two wedge-like inclined planes of steel, the car was gotten back with but little delay; but the poor animal was found with her neck wedged between the wheels of the following car. After trying in vain for fifteen minutes to back or pull the rest of the train over the body, they concluded to take an axe and cut off her head, after which she was pulled out, loaded up on a flat, and we went ahead.

A few miles below Honda, the mountains, which here are barren, dusty, precipitous, and furrowed with gullies and ravines, close in on the river until it is shut in in a deep gorge. At Honda, there flows into the Magdalena from the west the Gualí, a small, swift, and extremely muddy stream of some thirty yards in width; and a few hundred yards above, a second and smaller stream comes in. Between these there is a comparatively level terrace which widens considerably as one goes back from the river, and on this and along the river-shore the town is built.

We reached Honda about five, and went at once to the best hotel, a very neat one kept by two Englishmen, Messrs. Bowden and Willcox. It was a positive luxury, after being cramped up on the steamer for so many days, to get into a clean and spacious room, to find cots with clean sheets, and above all to have clean and appetizing food. After seeing that Alice was comfortably fixed, Cabell and I went out to call upon our consul, Mr. Henry Hallam, and to take a look at the town. We did not find Mr. Hallam, but at his office was a cablegram, sent from New York the preceding day, saying that all were well at home.

The town is not of much size, and offers nothing of especial interest. It is said to be the hottest place on the river, and deserves its reputation. It is shut in by the parched and baked mountains, and the few breezes that stir feel like blasts of hot air from a furnace. The houses are of the usual type, some thatched, some tiled. Through the enterprise of Mr. Hallam, water has lately been brought into the town. This gentleman has also established a line of wagons running westward to Mariquita over the terraces of the valley of the Gualí. I mention this as wheeled vehicles are practically unknown throughout the interior of Colombia. I was told that the muddiness of the Gualí was due to the hydraulic working of gold mines near its head-waters. This river was in former times spanned near its mouth by a ponderous masonry bridge of two arches, but this was destroyed by the earthquake of 1805, and now there is a fair iron bridge thrown across from the old abutments, and a short distance above there is a second bridge of wood. In the upper members of this iron bridge several large swallows had their nests. The centre pier of the original bridge remains, twisted to one side, and leaning up-stream. There are in the

RUINS OF BRIDGE OVER THE GUALÍ DESTROYED BY EARTHQUAKE.

town the ruins of a large cathedral which was destroyed at the same time.

The Magdalena here is very swift, the rapids in front of the town being like those below Niagara Falls, and it is of course impassable for steamers; but above the rapids there are some small steamers, running irregularly, which have at times continued the navigation of the river almost, if not quite, as far as the town of Neiva.

We saw piled up near the railroad station many small bags filled with a heavy sand-like silver ore, intended for shipment to England. Along the streets I saw a number of men with bad-looking ulcers about their ankles and shins, and a few with elephantiasis, a form of leprosy in which the ankle thickens enormously.

We were so pleased with our hotel that we thought of waiting here for several days to recuperate, but about dusk Lindauer came in to say that he would leave for Bogotá early the next morning, and that his muleteer had enough mules to supply us also, so we concluded to go on, and accordingly sent our trunks on ahead, so that they could be gotten across the river before we started. It was clear and hot.

Sunday, July 3, 1892. For the last five or six days on the river we had been without ice, and for a refreshing drink had taken a great deal of lemonade made from the limes, or "limones," that were found in abundance at every village. This had somewhat upset me, so I was not feeling partieularly well; however, we had a light breakfast at six, and started soon after. There was no train running, so we had to walk up to the ferry at Arranca Plumas, about a mile above the town. It was the ordinary swing ferry; a

SWING FERRY AT ARRANCA PLUMAS.
(After Millican.)

wire cable is stretched across the river, and on this a pulley runs. The boat, a large flat lighter, is fastened diagonally to the pulley, and the force of the current carries it across. It usually stops about twenty feet from the shore, and is hauled in the rest of the way by a rope thrown out from the landing. Once across, we scrambled up a steep and rough bank of loose pebbles and sand to a little ledge some thirty feet up the mountain-side, where there were four or five miserable bamboo and thatch huts. These, although their thresholds were on a level with the road, were thirty feet from the ground at the back, and supported on rickety bamboo poles. The floors were of split bamboo with cracks through which one's foot might easily slip. In these huts were sold various drinks and some dirty food for those whom hunger compelled to eat there. In front of them were great heaps of boxes and bales on their way to the interior. This is the terminus of the high-road to Bogotá, a city that now claims over one hundred thousand inhabitants.

Of course, our trunks had not gotten across after all, and when they were finally over, the mules had not arrived, and when the mules came, we were two hours in loading. Whilst waiting here, we took a poor breakfast to fortify ourselves for the road ahead of us. In the trees just at the landing I saw several large flocks of the orange-chinned parrakeets.

Our trunks were lashed with ropes of raw hide, one on each side of the little mules, and smaller parcels were put between. If the trunks did not balance, the lighter one was made heavier by tying stones to it. The mules have no other harness than a pair of pillow-like pads, which are furnished with both breast-straps and breeching. When all are loaded, they are started off by the drivers, or "arrieros," who follow on foot, keep the herd moving, and drive in the stragglers. The arrieros keep up a continual whooping and whistling, so that the mules may know that they are close behind, applying to them a choice selection of epithets, — "animalito," "mula del diabolo," etc. The loads are continually slipping, and when they slip must be rearranged at once. The arrieros are very dexterous at this. They throw their poncho over the mule's head, to blindfold it, and it stands perfectly quiet until the poncho is removed. They go along at a pretty good rate, but it is pitiful to see the little creatures staggering under two enormous packing-boxes as large as themselves. Often, when they get a chance

PACK-MULE WITH TRUNKS AND SLEEPING-MATS.

to stop, they lie down at once, and then cannot rise without the help of the arriero, who is certain to add blows to his aid. At numbers of places along the road we saw bones where the poor animals had died on the way. In this manner all freight is carried to and from Bogotá. We met a great many trains on their way down to the river. Some came unloaded, to carry back freight, but the greater part brought down bales of hides or bags of coffee.

We finally mounted and started off shortly after eleven, leaving the baggage to follow on. Alice and I rode horses; the rest were mounted on mules. The saddle, bridle, etc., are spoken of collectively in Spanish as " la montura." Our saddles had large horns, and were furnished with breast-straps as well as with both crupper and breeching. The bridles and bits were very heavy, the stirrups of brass and shaped like a Turkish slipper. The men, when riding, wear enormous spurs and a kind of leggings called "zamorras," something like the baggy rubber leggings used among us. They are made of canvas, rubber-cloth, or of leather, and are buckled together at the waist, thus forming a pair of trousers without a seat. Some that I saw were made of puma-skins. They are so voluminous that they completely cover the rider's feet, and when he dismounts they look like an awkward skirt and interfere with

ADJUSTING LOAD ON PACK-MULE.

his walking. (See page 97.) For the first two miles the road, ascending slowly, ran along the river to the south over what was once the beginning of a railroad. The embankments had washed away in many places, the cuts had caved in, and at one spot we passed a dilapidated old locomotive rotting away, with weeds growing over the boiler. This road was to have reached Bogotá, but the funds gave out with the first two miles. At the end of this we turned in abruptly to our left and began a steep ascent, zigzagging in and out of the gulley-like ravines that ran down to the river. When near the crest of the first ridge, the road ran over a rocky surface which seemed to me impassable. It sloped up at an angle of about forty-five degrees, but the feet of the mules had worn little pocket-like steps in the stone, and our animals went up without a slip. At the top we went through a narrow gorge, then along over comparatively level ground for a short distance, then up and through a second gorge so narrow that my stirrups scraped the sides, and down and across a rough valley several miles wide. This valley was hot and dry, but in the centre we crossed quite a large stream flowing to the south, and on the farther side we followed up the partly dry bed of another watercourse until we struck the foot of the first heavy range. Here the worst part of the road began.

All travelers in Colombia, from the time of Humboldt to the present day, have commented upon this road from Honda to Bogotá, and all agree in calling it superlatively bad ; but none have done it justice. In my limited experience I had been over some of the worst roads in the western part of North Carolina and in West Virginia, and I could not conceive that roads could be worse, but they are pleasant drives compared to this. I am powerless to describe it, and the photographs which I took on my return trip give no idea of the steepness of the road, since I had to point my camera either uphill or downhill, and thus the perspective of the slope was lost. In former times this road had been paved with blocks of stone, some of them as large as pillows. This pavement was in some places intact, but in a great many places it had been destroyed. To get a faint idea of the unpaved portion, conceive the dried-up bed of a rocky stream, filled with stones from the size of a barrel down, placed upon a hillside with a slope as steep as a roof. The paved parts were even worse on account of the slippery foothold that they

afforded our animals.   On the opposite page is an alleged view of a
portion of this road, but I will venture to say that the artist was
never in Colombia, or never saw even a photograph of this road.

PORTION OF PAVED ROAD TO BOGOTÁ.

I have introduced it simply to show what is the generally accepted
idea of South American roads.   The cut on page 241 of Mr. William E. Curtis's work, on " The Capitals of Spanish America," is
much more like the true state of the case.   The road went up the

ROAD TO BOGOTÁ.
(By permission of Bureau of American Republics.)

almost perpendicular crests of the foothills, zigzagging back and
forth at every ten yards, the pavement being built in steps up which
the poor mules toiled.   After about three hours' climbing, we
stopped for rest at Las Cruces, a mud and thatch inn on the right
of the road.   We found the air here decidedly cooler.   Here I got
some good oranges, and some green cocoanuts which were not nearly
so good as those that we had found at Barranquilla.   The country
through which we had passed to this point was parched and in some
places almost barren, being covered with a coarse grass and cactus;
but farther on we struck the forest, and found little cool streams
crossing the road, and everything was fresher.   I saw in the valley
many beautiful butterflies (some morphos especially being of large
size and brilliant color), a few humming-birds, and several flocks of
the blue-rumped parrakeets.   After about three quarters of an hour's
rest, we started again, and found the road growing steadily steeper
and worse, and shortly after four o'clock we stopped at a second
inn, Consuelo (consolation), where we concluded to spend the
night.   We were still half an hour from the summit, with the worst
of the road ahead of us; but although we had traveled only five
hours, we all felt somewhat used up, partly on account of the heat
and partly because of the roughness of the road.   The view from
this place was magnificent.   We were up between five and six
thousand feet, and could see across the valley of the Magdalena
to the distant range of the Cauca.   We found the air and water
much cooler, and needed blankets at night.   Alice and I were given
a little room in which were two wooden frames with cowhides
stretched over them for beds.   These we found to be swarming
with fleas, bedbugs, and a kind of flying roach an inch and a half
long, so we spent a wakeful night, tormented by bites.   The rest
of our party were given cots in the main room.

The landlord, Don Clemente Mejija, kept a blank book, by way
of hotel register, in which his various guests had indulged in their
fondness for poetry by writing, above their names, verses in praise
of the host and of his hospitality, or by giving vent to the emotions

inspired by the sublimity and beauty of the view of the distant mountain ranges.

In the yard in rear of the house was chained a long-tailed monkey, black with a white face, and there was also a cage of dull colored thrushes, marked somewhat like a newly fledged robin, but not quite so large. Don Clemente had a tame troupial which was allowed perfect liberty, but which came from the forest when called.

On the road we passed many peons bent under heavy loads of

ON THE ROAD TO GUADUAS.

over one hundred pounds, the weight being supported partly on their shoulders and partly by a strap passed across their foreheads. It was clear and hot.

Monday, July 4, 1892. We were up early this morning. As I was feeling worse, we decided to go on only as far as the next town,

Guaduas, and stop there, but as Lindauer was going to push ahead, he said good-by to us, and hurried on. We had a light breakfast, and started off about eight. Alice was very nervous about the road, and walked a good part of the way to the summit and down the other side. We reached the crest about nine, going up some places worse than a staircase, and just before reaching the top, through a deep and crooked gorge not wide enough for two animals to pass. I saw here the use of the brass slipper-shaped stirrups. In turning sharp angles, my feet were often pressed against the stones at my sides, and without these stirrups the barefooted riders would have their feet injured. We rode along the ridge for a few yards, and then began the descent.

"A DEEP AND CROOKED GORGE."

At one place the crest was barely ten feet wide, and fell off abruptly on each side for several hundred feet. From this point the view was grand. Through the clouds across to the west we caught glimpses of the perpetual snow on the Peak of Tolima and the snow fields of the Páramo del Ruis. To our left, to the southeast, lay Guaduas in the valley below us. It looked very near, but we were two and a half hours in reaching it. We went obliquely down the side of the mountain, and found the road not so bad as on the other side except at one place near the foot of the descent, where it ran over a hard stone lying in strata, which sloped in the same direction as the surface of the soil, so it was like riding along on a roof with no foothold for our animals.

Alice, in her nervousness from loss of sleep and from thinking about the road ahead of us, had not eaten anything before leaving Consuelo, and was now feeling faint from hunger, so we stopped at an inn at the foot of the mountain, and tried to get something to eat. I asked in succession for eggs, bread, coffee, plantains, rice, etc., until I had exhausted my vocabulary, but received the same

ROADSIDE INN NEAR GUADUAS.

answer to all my requests, "No hay" (there is none), so we had to push on.

From this point for about two miles the road ran over comparatively level ground, crossing two little streams on the way. The land was cultivated in places, and there were on either side of the road a number of little huts surrounded by small groves of orange-

PLAZA AND CATHEDRAL AT GUADUAS.

trees, coffee plants, and plantains. As we entered the town, the
road became a narrow paved street with a gutter of running water
in the centre, and just as the land began to rise to meet the second
range of mountains, we came out into the principal square, the Plaza
de la Constitución.   This was a large paved square with a fountain
in the centre.   On the eastern side was the cathedral, and on the
three remaining sides were various stores and public buildings, the
greater part of them of two stories in height.   About the centre of
the row of houses on the northern side was the only hotel in the
place.   It was of two stories, facing the plaza, the lower front rooms
being used as a store, and the one large room above as the recep-
tion or sitting-room.   Back of this was a square courtyard, and
farther back a second.   The rear of the house overhung a swift
running brook.   The entrance was through a narrow passageway

which was paved with small brown and black cobblestones arranged
in a very graceful arabesque pattern. This opened into the first
court, whence a staircase led up to the second floor. All of the
back rooms on the lower floor were used as storerooms and stables,
and above were the bedrooms. The dining-room was in the portion
separating the two courts. It was with a sensation of great relief
that we rode in through the passageway and dismounted. Upon

OUR HOTEL AT GUADUAS, FROM THE PLAZA.

asking for the proprietor, we found that he and his wife had gone
off to take a bath in some stream near the town, and they did not
return until towards sundown. There was an entire lack of system
and order in the house, and things seemed just to run themselves,
but after a while we managed to get some rooms, and in about an
hour and a half we had some eggs and coffee. Our rooms were
perfectly plain, and with no other furniture than canvas cots. After
trying to rest awhile, Cabell and I went out for a short walk to look
around. We saw a good many birds, flycatchers, swallows, turkey-

buzzards, black vultures, anis, and flocks of the little blue-rumped parrakeets. Some of the swallows that we saw were somewhat like our purple martin, a little smaller and not so brightly colored, and they had their nests under the curved tiles of the roofs.

The town is larger than Honda and is spread out over comparatively level ground. The houses are of the usual type, though many are roofed with tiles instead of thatch. Everything seemed dull and sleepy except the cathedral. During our stay some traveling missionaries were visiting the place, and the church bells were jangling from morning till night, and crowds were going in and out all day long. The valley is fertile and the climate delightful, the temperature far cooler than at Honda, and blankets are needed at night. Guaduas is said to be about 3,400 feet above the sea.

GOITRE.

I was struck with the great numbers of women of the poorer class suffering from goitre. Hardly one in five of the middle-aged women was free from it, and many of the men were also sufferers. Some have attributed this disease to the drinking of water from melted snow of the snow-clad peaks, but hardly within a week's journey of Guaduas could such water be found. Others have attributed it to living at high altitudes, but there are many people living in higher regions than Guaduas who are not affected. In Guaduas I found that the women were more affected than the men, especially the women of the laboring class. The carrying of heavy burdens partly supported by a band passing across the forehead necessitates a tension in the muscles of the neck and throat which may have some influence in producing the enlargement. So accustomed to it are the people here that (I was told) they even regard the goitre as a mark of perfection, and those who do not have it are considered as departing from the normal.

I also saw many children with some of the nails missing from their toes, and was told that this was caused by neglecting to pick out the "chigoes," or "nigoes," as they are sometimes called. These little vermin burrow under the nail and deposit their eggs in a sac. This can be easily picked out with a needle, but if neglected until the eggs hatch they produce ugly sores, sometimes attended by loss of the nail.

When the proprietor finally returned, to our surprise we found that he was a Virginian, a Mr. David Bain, who had been out in Colombia for over twenty years, and who boasted of being even more indolent than the natives. Upon learning that we also were Virginians, he did all in his power to make us more comfortable, and gave Alice and myself the room over the entrance, which had the advantage of having a window facing on the plaza. The hotel was once a private dwelling, and must have belonged to a person of relatively considerable wealth. The floors were paved with a large coarse tile, as thick as our brick, but about ten inches square. These were warped, and had wide cracks between them which were strongholds for innumerable fleas. In going over the house, I noticed at one place, where the plaster had fallen off, that the laths, which were of split reeds, were fastened to the joists by being tied with a slender vine, and not nailed as ours are.

At night, as I was feeling no better, I found the doctor of the place, who gave me a prescription of ipecac, chalk, and opium. The night was cool, and we would have rested well except for the multitude of fleas that fairly devoured us. We spent a wakeful night and were fearfully bitten.

Tuesday, July 5, 1892. I was worse this morning, having slight symptoms of dysentery, so sent again for the doctor, who gave me some calomel, after which I kept on the bed all day and spent the time reading a Spanish edition of the "Scientific American." In the afternoon Cabell went out with his gun, and later returned with some birds, among which were six large humming-birds, all of the same species, but different from any that we had met so far. They were large, the males brilliant green above, the throat and breast black with an edging of deep blue, the tail a rich purple bronze, a white downy puff on the belly, and a white speck back of the eye. The female was similar, but below was white with a black band down the centre of the throat and breast (*Lampornis nigricollis*). These he had found feeding on the scarlet blossoms of a large tree near the town. He also brought in a green-naped pigeon, like the one that I had killed on the river, and a woodpecker about the size of our yellow-bellied, but colored somewhat like the red-bellied. Its head, throat, and below were ash-buff, the centre of the belly and back of the head washed with red, back closely barred with black and white, wings and two central tail-feathers black and white, remaining tail-feathers black with white tips, and rump white (*Centurus terricolor*).

It was clear and warm.

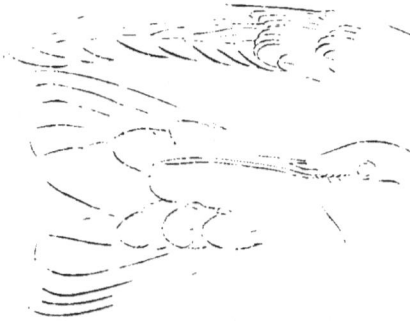

LAMPORNIS VIOLICAUDA.
(From Elliot.)

$\frac{4}{5}$

G.G.Keulemans, del.   Mintern Bros. imp. lith.

EUPSYCHORTYX LEUCOTIS, *Gould*.

White eared Partridge.

Wednesday, July 6, 1892. I felt a little better, but was still unable to eat anything, and stayed in bed all day.

Thursday, July 7, 1892. I felt worse in the morning and sent again for the doctor, but was told that he had gone away the day before and would be back in "about a week." I was feeling pretty blue over this news when my brother came in to say that an English gentleman stopping in the hotel, a Mr. George Child from Bogotá, on his way to Honda, learning of my sickness and of the fact that there was now no doctor in the town, had kindly offered me a remedy which he had with him, "chlorodyne." Within ten minutes after taking the first dose I began to feel better, and from this point I recovered rapidly. Cabell made some beef tea for me later in the day, which was very strengthening. In the afternoon he went out with his gun for a little while and returned with quite a collection of birds. Among them were a pair of cardinals, an ani like the one killed in Barranquilla, and a hawk rather smaller than our Cooper's hawk, beak horn-blue and black, eyes, feet, cere, and skin of face yellow, above plumbeous, the tail black barred, the rump white with black bars, the primaries chestnut, black barred, under coverts finely marked with chestnut arrows, below plumbeous turning to rusty, breast and belly closely barred, the bars growing smaller towards the vent, and thighs closely barred with rusty (*Rupornis magnirostris*). The natives called this a " garrapatero," or tick-eater, but they apply the same name to the milvago and to the anis. He also brought back a tanager of the usual size, and with a beak much like that of our summer redbird, uniform blue-black with white shoulders and under wing-coverts (*Tachyphonus melaleucus*), a little ground dove, a blue-rumped parrakeet, and a pair of partridges, both males, about the size and shape of our " bob-white." Their back, wings, and tail were very like those of our bird, top of head buffy and black, with a recurved crest of clay-colored feathers, chin, forehead, and ear-coverts whitish, throat, stripe above eye, and malar stripe rufous, breast mottled black, white, and rufous, the rufous prevailing on the lower breast (*Eupsychortyx leucotis*).

Towards evening I was feeling very much better, so I went in to the supper-table, though I confined myself to beef tea. I enjoyed conversation with Mr. Child, as he was well informed about the country. Whilst we were at the table, Mr. Millican, that energetic collector of orchids, came into the hotel. He was just on his way to Honda with a consignment of plants, which he was going to ship to England, and then return at once to his collecting-ground. About eleven o'clock that night I heard quite an uproar, and upon inquiry found that he had unfortunately uttered some criticism about the hotel, which reached the ears of the landlady, and she was so incensed that she immediately turned him and his servant out into the street, driving out his mules, and throwing their saddles out of the door after them.

It was clear and hot during the day, but delightfully cool at night. The fleas, however, entirely prevented our sleeping.

Friday, July 8, 1892. I felt much better, but still stayed in or near the hotel the greater part of the day, and confined myself to a beef-tea diet. At breakfast I thought that I would try a soft-boiled egg; but when I cracked it into my plate, it was not done, so I thought then that I would have it scrambled; and, to hurry it up, I took it out to the kitchen myself. When I had explained what I wanted to the cook, — a dirty old Indian, — she took my plate and scooped up the half-done egg in her hand, and transferred it thus to her pan; so I changed my mind about wanting egg after all. Speaking of this reminds me that in Guaduas, and in other places in Colombia, they call scrambled eggs " pericos," which means, literally, little parrots; but why they are thus called I could not find out. The kitchen of our hotel was peculiar. It was a large room, without fireplace, stove, or chimney. Along two sides ran a built-up ledge of stone, much like the hearth in a country blacksmith's shop. On this all the cooking was done, a dozen little fires being built at intervals. All of the earthenware utensils made in the country have round bottoms and no legs, so they cannot be made of themselves to stand upright, but three round cobblestones must be placed around the fire, and the vessel placed on them.

I thought it best to have my drinking-water boiled whilst I was sick, so purchased an earthenware jar for the purpose; but I had great trouble in the matter. At one time, as soon as the water boiled it was taken by the servants to wash dishes; at another time, when I asked about it, the cook, to see how hot the water was, put her hand into it.

I was also occupied for a portion of the day in trying to destroy

MARKET IN PLAZA AT GUADUAS.

some of the fleas in our room. I purchased a pound of crystallized carbolic acid, with which I made a strong solution, and scrubbed the floor with a broom, being careful to let the liquid run into the cracks; but at night we were bitten as severely as before. Every

morning our white blankets were found full of them. They creep
into the wool as they would were it growing on an animal's back.
The few dogs that I saw around Guaduas were abject-looking
creatures, and appeared as if life were a burden to them. The
most of them were hairless. They are not only devoured by fleas,

A PACK-OX AT GUADUAS.

but there are other vermin which burrow under the skin, like the
" wolf " in our rabbit. Cattle suffer in the same way, and we saw
some mules and horses with one ear gone, due to the attack of some
insect.

This was market-day, and the plaza was crowded. I walked
around to see what was going on, and to take some views with my
camera. There was the usual assortment of fruits and vegetables
for sale in the market, and nothing remarkable except that at one
place I saw unborn pigs exposed for sale. This, I thought, was

rather getting ahead of us in our dish, sucking pig. Salt, of which
the government has the monopoly, was weighed out in little scales
as carefully as a druggist weighs his medicines. The duty on salt
is about three and a half cents per pound, and in Guaduas it was
sold at ten cents per pound. Beef is very good here, and cattle
are butchered every morning. The hides, which are exported in
large quantities, are prepared by simply stretching them out with
pegs over the ground, hair side down, but clear by about ten inches.
When dry, they are folded up into squares about the size of a coffee-
sack, and then tied up into bales.

A good deal of the produce from the neighborhood was brought
in on the backs of bullocks. They are said to be even more
sure-footed than the mules, though slower. Such things as fruit,
vegetables, earthenware vessels, etc., are put into purse-like bags
of a coarse netting, and then loaded on the pack-animal. (See
page 88.)

In the afternoon Cabell went out with his gun, and later Alice
and I went out a short ways to meet him
on his return. He had been to some flow-
ering trees near a coffee plantation along
the road, and brought back eleven humming-birds of eight different species. They
were, first, a pair of the large black-
throats (*Lampornis nigricollis*). Second, a pair, male and female, but slightly
smaller; the male green above and below,
with broad, black tail-feathers and conspicuous white plume-like under tail-coverts; the female was similar, but had
more gray in the green below (*Hypuroptila buffoni*). Third, a pair, golden bronze
above with a greenish tinge, the central

HYPUROPTILA BUFFONI.
(From Elliot.)

tail-feathers the same; the others chestnut, with purplish bronze
and white tips, below gray, with a darker patch on the throat

(young of *Chrysolampis moschitus*). Fourth, a rufous-tailed humming-bird, like those that we had gotten on the river (*Amazilia
fuscicaudata*). Fifth, similar to the last, but tail blue-black and
forehead dirty blue (*Amazilia cyanifrons*). Sixth, small, brilliant
emerald green above and below, and tail blue-
black (*Chlorostilbon angustipennis*). Sev-
enth, similar in size and shape to the last, but
tail different, and belly very deep metallic
blue (*Damophila julia*). Eighth, very minute, greenish above, white collar and flanks,
below rufous, tail chestnut, with dark green
subterminal bar and a few metallic amethyst
feathers in a dusky throat (young of *Acestrura heliodori*). The total length of this
last was just two and a quarter inches. He also shot an owl about
the size of our short-eared owl, but with long ears, the plumage
more rusty than that of our long-eared owl and the feet more bare
(*Bubo mexicanus*). This he had found roosting in some thick
coffee plants.

DAMOPHILA JULIA.
(From Elliot.)

At night two friends of Mr. Bain came in and gave us some
excellent music. They played on two instruments shaped like
guitars, but both strung with four double strings like a mandolin.
The smallest, which was very small, was called a "tiple;" the
other, about the size of a guitar, was called a "bandola." The
performers excel in keeping time.

I saw to-day a new fruit, a "badea." It was the size and shape
of a small pumpkin, and when cut open its flesh made the resemblance stronger. In the cavity inside were many seeds, each one
surrounded by a pleasantly acid pulp, and this was the part that
was eaten. There are pomegranates in Guaduas; but they do not
come to the same perfection as at other places. I also saw here for
the first time another fruit, of which I had heard so much, and
expected to find delicious, the "granadilla," or little pomegranate,
and to my surprise recognized an old friend, the "may pop" of

our Southern States, the fruit of the common passion-flower, which is so abundant in our corn-fields in the early fall. It owes its Spanish name to the fact that, like the pomegranate, the portion that is eaten is the pulp around the seeds. It was my experience to be greatly disappointed in the fruits of the tropics; but as this disappointment was only in those fruits which I had never before tasted, it may be that in time my taste would have been educated up to the point of liking them. The oranges, pine-apples, and bananas were incomparably superior to any that we get, and I be-
came in time very fond of the níspero ; but after once tasting the mango I had an aver-
sion to it amounting to disgust, and were I to describe the flavors of the many other fruits that I tried, I would say that they varied from that of a pumpkin to that of our paw-
paw. There being no

TIRED OUT.

frost in this climate, some plants which regularly die every winter with us, grow here indefinitely. Such are the *Palma christi*, or castor-oil plant, which becomes a fair-sized tree, and cotton. I saw several cotton plants which might almost be called small trees; however, the bolls were very small, and produced an inferior cotton.

The principal forage for horses and mules is young sugar-cane, which is chopped up with a machete into little pieces of an inch in length. The animals are very fond of it; and it seems to suit them, for, notwithstanding their hard labor, they all look sleek and in good condition.

Saturday, July 3, 1892. Cabell, Alice, and myself were up early this morning, and, after having some eggs and coffee, went back along the Honda road for about a mile to a coffee plantation, a place called "Tuscola," where we had obtained permission to shoot, and where Cabell had found the humming-birds on the day before. When we reached the place, we found on the roadside a tree in bloom, and around this some humming-birds were feeding. The first one that I saw and shot proved to be a new one, but in poor plumage. It had a long, curved beak, the lower mandible yellow. Above it was dull green; below dusky, mark above eye and ear coverts black; a whitish superciliary streak. The two central tail-feathers were prolonged and pointed, their tips white and bases dusky (*Phæthornis superciliosus*). From here we went up into the plantation, which we found grown up into a perfect jungle. It consisted of coffee,

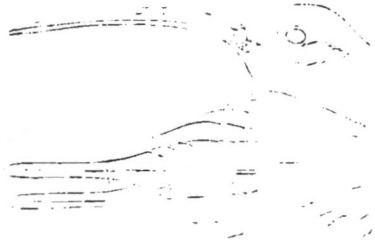

PHÆTHORNIS SUPERCILIOSUS.
(From Elliot.)

orange, and cocoa trees planted together; but they were all shaded by large forest-trees, and interwoven so thickly as to be almost impenetrable. We found a cool, shady spot near a little trickling spring, and Alice arranged herself comfortably with a book, whilst we hunted around within fifty yards. As we drew near some trees with a white, fringy blossom, we could see humming-birds darting about among them, and could hear their humming and buzzing at quite a distance. The first one that I shot was a perfect gem, — by far the prettiest that we had yet seen, — a male, ruby and topaz (*Chrysolampis moschitus*), in perfect plumage. Its body was brownish green; tail rich chestnut; head above ruby; gorget brilliant, golden yellow. Later, I killed two young males, similar to those that Cabell had shot the day before, but having a few ruby feathers on the crown and a few topaz ones on the throat. Though we saw great numbers of humming-birds, we had to select our shots carefully, for if one fell in the thick underbrush, it was a hopeless task to look for it. We found that we had either to wait until they were over an open space or else shoot them immediately overhead, so that they would drop at our feet, and even then we lost a good many.

CHRYSOLAMPIS MOSCHITUS.
(From Elliot.)

A third new kind that I killed was a small bird, green above, green and white below, chin, spot back of eye, and flanks white, and gorget amethyst. Its tail was peculiar, black, and forked, the two outer feathers on each side being reduced almost to a bare stem (*Acestrura mulsanti*).

In the grove we picked some of the most delicious oranges that I ever tasted. We started back to the hotel shortly after ten o'clock, and on our way stopped to fish in the little stream that we crossed. We used grasshoppers for bait, and in a few minutes caught a half dozen small fish, shad-like in general shape, but with the fleshy dorsal fin of a trout. Their jaws were also much heavier than those of a shad, and in the lower jaw in front were a pair of strong and sharp teeth (*Characin sp.*). We broke the only hook that we had with us, so had to stop fishing.

ACESTRURA MULSANTI.
(From Elliot.)

In the afternoon Cabell and myself returned to the coffee plantation, and got seven or eight humming-birds. It was close cloudy at this time, and the light under the trees was barely more than twilight, so we lost more of the humming-birds than we got. I myself lost nine. We got in all to-day twenty-one, among which were five males of the ruby and topaz, all in fine plumage. In the morning I saw flying over high in the air a pair of fork-tailed flycatchers (*Milvulus tyrannus*); but they kept on out of sight without lighting. We saw a small flock of partridges; but although we ran at once to the spot where they lit, we did not succeed in flushing a single one.

I noticed all through the coffee plantation a number of little beaten paths, from two to three inches wide, and perfectly cleaned of all grass, leaves, twigs, and even small gravel. They looked like the impressions left on a grass lawn when a piece of timber that has been lying on the grass for six months or more is taken up. A peculiarity of these paths was that even when they passed for several yards over the bare surface of the out-cropping stone, they could still be plainly traced, for the lichens and dust had been cleaned off until it looked as if an attempt had been made to polish the stone, and the path was lighter colored

SAÜBA OR LEAF-CARRYING ANT. — 1. WORKER-MINOR;
2. WORKER-MAJOR; 3. SUBTERRANEAN-WORKER.
(From "The Naturalist on the Amazon.")

than the adjacent surface. I was wondering what animal had made these, when I came upon one in use. Thousands of ants were hurrying along in opposite directions, those going in one direction being empty handed (or rather, empty jawed), each one of the others carrying held up edgewise a piece of leaf, approximately circular in outline, and about the size of one's finger-nail. The ants were a little smaller than our large black wood-ant. They were the leaf-cutting ant, described by Bates in his " Naturalist on the Amazon." I later found their hill. It was about a foot in height, but certainly twenty feet in circumference. There were numerous entrances, and their highways radiated in every direction. I followed one for about half a mile. The leaves came mainly from the coffee plants.

An interesting plant here is the guadua (*Guadua latifolia*), from which the place derives its name. This is what I have spoken of as the bamboo. It grows in graceful feathery clumps and reaches a large size. I saw some nearly fifty feet in height and as thick as a man's thigh. It sprouts up like an asparagus plant, that is, shoots up a large, club-like growth which does not put out leaves or branches until it reaches a good height. It has a hundred uses; many utensils and vessels are made of the joints, and it is one of the most universal building and fencing materials. The smaller ones make good fishing-poles.

At the supper-table we met the first ill-mannered person whom we had thus far encountered in Colombia, and I am ashamed to have to admit that he was an American drummer.

We spent another wakeful night tormented by the fleas. It was clear and warm in the morning, but close cloudy in the afternoon.

Sunday, July 10, 1892. Alice, Cabell, and myself went out again early this morning to the same place to which we had gone the day before, but we came back soon. We got twelve humming-birds, but there were no new ones among them. They were divided among the following species: five emerald green ones (*Chlorostilbon angustipennis*), two of the small emerald green and blue ones

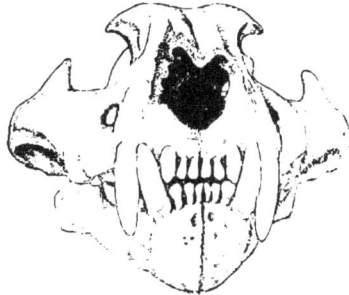

JAGUAR SKULL.

(*Damophila julia*), one of the smallest kind (*Acestrura heliodori*), one with an amethyst gorget (*Acestrura mulsanti*), two young ruby and topaz, and one of the large black-throated ones (*Lampornis nigricollis*).

In the afternoon we were busy getting our things together, as we concluded to start back on the following day. The climate of

JAGUAR.

Guaduas is delightful, and the place promises well, but Alice is getting worn out by loss of sleep and by poor food. I arranged about our mules and two peons to accompany us and bring the mules back. Our specimens so filled a trunk now that I had to purchase another pataca. In the little shop where I bought the pataca, I saw several deerskins, apparently of just the same color as our Virginia deer, the fawns being likewise spotted, and the skull of a young jaguar that had been killed not very far from Guaduas. The shopkeeper, seeing that I was interested in it, insisted upon making me a present of it. The native name for jaguar is "tigre." They are said to be especially abundant in some portions of the Magdalena Valley. They are heavier and more stocky than a leopard, but otherwise are much like that animal. There are also many pumas in Colombia, and I saw numbers of their skins. Mr. Bain wore a handsome pair of zamorras made from a pair of puma-skins which had come from near the snow line of the Páramo del Ruis. The natives call the puma "tigre colorado," red tiger, or sometimes "león," lion.

The religious ceremonies which had been going on all during our stay culminated late in the afternoon with a procession around the plaza. Some really pretty arches were erected at the four

RELIGIOUS PROCESSION AT GUADUAS.

corners of the square and covered with palms and flowers. The procession was formed on the steps of the cathedral and filed slowly around under the arches, halting at each one whilst one of the priests delivered a brief sermon. The column was headed by four musicians, these were followed by several priests, then came the communicants, little girls from five to ten years old and of all colors. They were dressed mainly in white, some with little gauze wings, as if to represent fairies at a fancy dress ball, and all wore flowers in the hair. Their mothers marched on either side, all dressed in black with a black shawl over their heads, and bearing a candle. After the girls came the little boys carrying small banners, and then came groups of men bearing on their shoulders platforms with wax figures of the Virgin, Saint Joseph, and other saints. The Virgin wore a crown and dress like those worn by Queen Anne. One of the saints was dressed like Charles the First. The houses facing the procession were made gay by flags and lace curtains draped over the balconies.

During our stay in Guaduas several detachments of soldiers passed through the town escorting government stores from the river to Bogotá. They usually rested a day in the town, and spent their time whilst there sitting in the shade of some doorway and playing cards on a poncho spread on the ground. They were armed with Remington rifles, but apparently knew nothing of keeping their weapons in order, for such of their pieces as I examined looked as if sandpaper and fat pork were the cleaning materials.

I approached a party playing cards on the hotel stairs, and picking up a cartridge-belt examined the cartridges. The bullets had all been drawn, the powder sold, and the bullets then put back. In some cases the bullet had been lost, but a wooden plug answered every purpose.

CHAPTER VI.

BACK TO BARRANQUILLA.

MONDAY, July 11, 1892. We were up very early, but owing to the usual delays did not get off until half past seven. Mr. Bain insisted upon accompanying us for a portion of the way. Besides our riding mules, we had three baggage mules, so we made quite a train. The air was fresh and cool, and we

"ALICE . . . DISMOUNTED ONLY FOR THE BAD PORTION ABOVE CONSUELO."

passed without trouble the bad spots in the road, so reached the summit in good time. Alice was more accustomed to the road by this time, so she dismounted only for the bad portion above

Consuelo. I carried my camera under my arm, and took various views as we went along. It was with feelings of regret that I took the last backward look at Guaduas, as we turned to go over the crest. Before us stretched a magnificent view. The valley beneath us was filled with clouds, but above them we saw the glittering snow of Tolima and of the Páramo del Ruis. I tried several views from this point with my camera, but much to my disappointment, when the plates were developed, the blue sky and the white peaks both came out white, and there was no contrast between the two.

At Consuelo we said good-by to Mr. Bain, and waved an acknowledgment to Don Clemente's "feliz viaje." We traveled along comfortably as far as Las Cruces, where we stopped for breakfast and to rest. For about twenty-five cents we got some bread, rice, and eggs, all nicely cooked, and some of the most delicious coffee that I ever tasted. From this point, as we descended, the heat increased until it became almost unbearable; however, as we wished to reach Honda before night, we had to push on, as the ferry stopped running at six. The distance from Guaduas to Honda is some sixteen miles, and we were traveling for about seven and a half hours. The latter part of the road was made doubly disagreeable by thick clouds of a suffocating dust in which our mules sank to their ankles at every step. We had no drinking water along the road, and all suffered from heat more or less. After an irritating delay at the ferry we finally got across, and about five o'clock I was relieved by once more reaching in safety Mr. Bowden's welcome hotel. Mr. Child, who had preceded us from Guaduas by several days, had engaged good rooms for us, so we were soon comfortably fixed. After supper Cabell and myself took a short walk through the town, and I purchased for fifty cents a very pretty tiger-cat's skin.

Shortly after passing Las Cruces, as we were riding along through a parched bit of scrubby woods, I heard a loud rustling noise as if a high wind were approaching; but in a short while I discovered that the noise was made by an immense swarm of grasshoppers creeping and hopping over the dry leaves. They were of comparatively large size, yellow with a black stripe down the back, and were wingless, that is, their wings were not yet developed. The noise that they made could be heard at least one hundred yards. I have heard in Florida a similar noise made by multitudes of fiddler crabs running over the dry marshes, but the noise of these grasshoppers was greater.

At another point on the road we passed a man who was carrying with him a half dozen game-cocks. These were arranged in a peculiar way, so as to be carried without injury. Each was furnished with a pair of trousers of cotton cloth into the legs of which its legs were thrust. The part corresponding to the seat was brought up and buttoned over the back, securing the wings, and on the back a loop was sewn by which the cocks were suspended from the saddle. They were thus carried in a natural position without chafing and without being able to strike at one another. It was clear and hot.

Tuesday, July 12, 1892. We stayed around the hotel in the morning, and later went for a short walk through the town, calling on our way upon Mr. Hallam. As we came back I bought for five dollars at a saddler's shop a very large and handsome jaguar-skin. Late in the afternoon Cabell and I took our guns and went out about two miles on the table-land in rear of the town. We saw very few birds. The land is hot and dry, with scanty vegetation, and promises little. On our way out an iguana about three feet long ran across the street just in front of us, and scrambled up among some vines on an old stone wall. Cabell shot a rufous ground dove (*Columbigallina rufipennis*) of the same kind as the one that we obtained in Barranquilla. We saw some partridges, kingfishers (*C. amazona*), flycatchers, anis, and some very small finches, but hardly anything else. We saw several small swarms of the same kind of grasshoppers that we had seen on the day before. The anis were feeding on them. There are some beautifully colored grasshoppers in Colombia. One that I saw several times was an almost metallic green with bright scarlet eyes, legs, and bead-like markings, looking like a jeweler's design in emeralds and rubies.

In the neighborhood of Honda there is an upright cactus which is used for hedges and fences, but which differs from the one used for that purpose in Curaçao. The latter is approximately circular in cross section, but the former is star-shaped, that is, it has a small circular core but wide radiating flanges, thus being very rigid and at the same time light. Near Honda I also saw a cactus much like the prickly pear, but its lobes were perfectly smooth and devoid of thorns. It was clear and intensely hot.

Wednesday, July 13, 1892. We stayed around the hotel the greater part of the day, and did nothing in particular. In the afternoon, Cabell and myself walked down to the river and watched a man fish for a while, but he caught nothing. We saw in an Indian's hut a domesticated bird called a "guacharaca." It was about as large as our ruffed grouse, but had much longer tail, legs, and neck, and a little head like a turkey's. Its plumage was dark without distinctive markings, and it had a slight gular pouch. This was an immature bird. The name "guacharaca," given to this bird from its call, recalls at once the Mexican bird similarly named, the "chachalaca." It was clear and hot.

Thursday, July 14, 1892. Cabell and myself went out about six this morning to the place where we went on Tuesday. We had hardly reached the spot when we saw several fork-tailed flycatchers (*Milvulus tyrannus*), and I shot a pair, male and female. Their bodies are about the size and color of that of our kingbird, light gray above, the head blackish with a concealed yellow patch. Their flight is so graceful that they seem to float through the air. They perch on the tops of small bushes, just as our kingbird does. A little later Cabell shot a new dove, a male, about the size of our Carolina dove, but with a short tail, reddish, the under tail-coverts reddish, and two blue-black streaks on each side of the head (*Zenaida ruficauda*, Bonaparte). He also shot a pair of the little

FORK-TAILED FLYCATCHER (MILVULUS TYRANNUS).

ground doves, and later I killed a second one of the rufous-tailed doves, also a male. We saw a covey of eight partridges; but although we ran in upon them at once, we flushed but one and did not get it. We were much troubled to-day by a sort of nettle which is very abundant here. It has large leaves covered with a multitude of hair-like thorns, which prick at the least touch and produce a burning pain which lasts for some time, and, to say the least, is very disagreeable. The heat soon became so oppressive that shortly after eight we turned back and reached the hotel about nine. After breakfast I skinned the birds, and the remainder of the day we spent around the hotel. An Indian boy brought to the hotel for sale a fresh fish of the same kind as the small ones that we had caught at Guaduas, but this one was about four pounds in weight. It had been caught in the river in front of the town.

We received some letters to-day, the first that had reached us since our departure. It was clear and hot.

Friday, July 15, 1892. We stayed around the hotel all day, saw about our tickets for the steamer America that was to go down the river next day, and got together our baggage. At Mr. Hallam's we had the pleasure of meeting the captain of our boat, Captain Bradford. This gentleman, a Georgian by birth, a graduate of the Naval Academy, and an officer of our navy, left the service of the United States to side with the Confederacy at the outbreak of the late civil war, and upon its termination settled in Colombia, where for the last twenty-five years he has run upon the Magdalena River. He is a gentleman of the old school, and we found in him at all times that courteousness which is so delightful, but which we now, unfortunately, so rarely meet. I should advise any future travelers on the Magdalena to make inquiries as to when the America will go up or down, and, if possible, take passage on her.

It was clear and intensely hot.

Saturday, July 16, 1892. We were up early this morning, and soon had our trunks packed, after which Cabell and I went out to get the tickets which we had engaged the day before. The fare for the down trip, owing to the shorter time required, is only two thirds of the fare up. We saw a very large iguana in one of the trees overhanging the Gualí at the old bridge. In the same tree I saw a new bird, a woodpecker about the size of our red-headed woodpecker, but with its colors distributed like those of Lewis's woodpecker. Its general color was a dark sage-green, its cheeks white, the back of its head red. I did not see its under parts. This bird I have not yet been able to identify (*Chloronerpes sp.*).

We had our breakfast about half past eleven, and afterwards went down to the station to take the train which was to leave at one. We reached the station about twelve minutes ahead of time; but the agent had gone off to his breakfast, and the conductor positively refused to allow me to put my baggage on board. I tried every argument, offered him money, represented to him that we had purchased our tickets for the steamer which was to sail at three o'clock, and that there was no other train until the next day, but my trouble was for nothing. At last I was so incensed that I did what I should have done at the outset, that is, I began to put the baggage on board myself; but one of the loafing officials, seeing that I had nearly everything on, gave the signal to start, and the train pulled out about three minutes ahead of time. I hardly knew what to do; but as Alice and Cabell were on board, I jumped on also, and called out to some one on the platform to look after the trunks that were left. We were an hour and a half in reaching Yeguas, and by that time I had cooled down a little. From here I telegraphed back to Honda, and finally got orders for the engine to return for my trunks, and Captain Bradford secured permission to hold his boat for me. I went back with the engine and flat car in thirty-two minutes, and returned in twenty-nine. The road was so rickety that I was in constant dread that we would jump the track. For this special engine I had to pay thirty-five dollars in paper.

As soon as I reached Yeguas our trunks were put on board, and the steamer started. We went up the river first for about a quarter of a mile, then turned and came down, running with the current like an express train. Captain Bradford had selected very comfortable staterooms for us, and did everything in his power to make our trip a pleasant one. The America is one of the mail

steamers, and is run on quite a different plan from the Enrique.
We made very few stops, and did not take on wood until we tied
up for the night.   Just after sunset we tied up, and as soon as the
gang-plank was put out, Cabell and I hurried ashore.   Within
twenty yards of the landing Cabell shot a large hawk which dropped
in a thicket near by, and as he ran to pick it up a large bird sailed
out of the forest, and lit in the tree over his head.   He fired at it,
and it spread its wings and glided down to the ground about fifty
yards off, where one of the little Indian boys with us ran and
brought it back.   Darkness comes on almost instantaneously here,
for in the minute or two that had elapsed since the hawk was shot,
it had become so dark that we had to give up the search for it, and
return to the boat.   We found Captain Bradford anxious on our
account, fearing that we would be snake-bitten.   When I came to
examine the bird that the little Indian had picked up, I found it to
be something new and very curious.   It was of the whippoorwill
family, but very large, measuring twenty-one inches in length by
forty-two in extent.   Its mouth, which was not provided with bristles,
was so large that I easily put a moderate-sized orange in it.   Its eyes
were very large and dark, the soles of its feet broad and flat like the
palm of a hand.   Its upper mandible had a tooth-like projection on
each side, and fitted over the lower.   Its tail was large and rounded,
and, like the rest of its plumage, was beautifully mottled with gray
and black.   Its back was rusty in places and its shoulders were
dark brown (*Nyctibius grandis*).   I skinned it the following morning.   It was a female, and had been feeding on large black beetles.
It was larger than a short-eared owl.

I bought at this place a very prettily marked tiger-cat's skin,
quite fresh, and saw several peccary-skins and a portion of the skin
of a tapir.   I was told that there were two species of tapir found
near here, one in the river valley and another on the mountains.

Above Yeguas I saw the same kinds of sparrow-hawks and rufous-winged buzzards that I had seen when here two weeks ago.
Below Yeguas I saw many blue and yellow, and blue and scarlet

macaws, parrots, hawks, and cocoi herons. For nearly a mile the steamer passed through an immense swarm of grasshoppers. They almost darkened the air, and actually bent down the bushes upon which they settled. They were of the same kind as those that we had seen at Honda, but were fully grown.

I had been told that no alligators were found above Yeguas, but when I went back on the engine for our trunks I saw several large ones within a couple of miles of Honda.

At night we were not troubled with mosquitoes, but made the acquaintance of a new insect pest, the "egén." This is a minute fly that causes a blood-blister the size of a pin's head to rise on the skin. It does not itch, so is not as irritating as the bite of a mosquito, but leaves a mark that lasts for a week or more.

It was clear and very hot.

Sunday, July 17, 1892. We started at early dawn, made very few stops during the day, and ran along rapidly. The river was

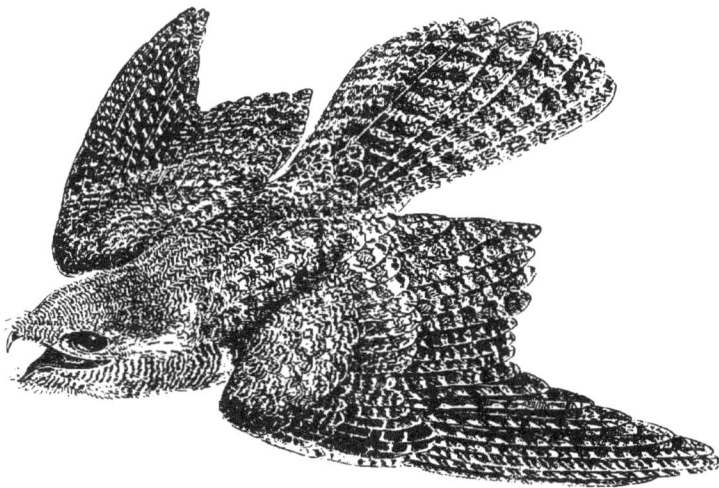

GRAND POTOO (NYCTIBIUS GRANDIS).

much lower than when we came up. We saw quantities of alligators, and shot at them many times. We struck a number, but
killed very few. Cabell killed a tremendous one, and we wished
very much that we could have gotten it. It was fully 250 yards
distant, and he killed it with the 32 Winchester after firing four shots
at it. Captain Bradford, who is an excellent shot, joined us in the
shooting. He used a 44-calibre Remington, which was much more
effective than the light Winchester. About noon we made a short
stop at Puerto Berrío. I hurried ashore with my gun, and in a
few moments shot a new toucan. It was similar in size and coloration to the one killed by Mr. Lindauer on the up trip. The beak
of this one, however, was plain, not serrate, and was chocolate-
brown, almost black below, greenish yellow on top, becoming pure
yellow at the tip. The skin of its face was a bright lemon-green,
feet lead-blue. The colors of the plumage were like those of the
first one, except that the rump was white (*Ramphastos ambiguus*).
This was a male, and was one of a large scattered flock.

We did not tie up for the night until it was too late to go ashore.
I saw during the day some capybaras, two kinds of macaws, some
guacharacas, and four kinds of toucans, the three of which we have
obtained specimens and a fourth whose under parts seemed largely
red.

The steward bought for the table a large turtle, or rather terrapin.
It had a smooth shell, a uniformly colored skin, a sharp pointed and
snake-like head with the eyes much nearer the tip of the nose than
in our river terrapins. The poor reptile was secured by having its
feet sewn together. It was clear and hot and rained at night.

Monday, July 18, 1892. We started very early, and made a good
day's run, passing Bodega Central, Puerto Nacional, and other
places, but made no stop long enough to go ashore for birds.
We shot a great many times at alligators. I saw several iguanas
at the mouth of the river Lebrija. At Puerto Nacional there was
one on the river's bank, and getting between it and the trees, I
made an attempt to catch it; but without the slightest hesitation

KING VULTURE.
(From " Riverside Natural History," by permission of Houghton, Mifflin & Co.)

it dived boldly into the water, and swam off beneath the surface as
easily as a frog. We saw several small bands of monkeys in the
trees as we passed along. After we had tied up for the night, the
mosquitoes became very troublesome. It was clear and hot.

Tuesday, July 19, 1892. We passed Banco early in the morning, and later the mouth of the Cauca, soon after which we made
a short stop at Magangué. The water was now too low for us to
get into Mompos. I saw during the day several herds of capybaras.

Before I got up, Cabell had seen a king vulture, and later in the day I saw a pair perched in a low dead tree growing in a marsh. At the distance that I saw them, they looked black and white with red heads. I also saw another new bird, apparently an ibis, very large, snowy white plumage with black head and legs (*Mycteria americana*). I saw large flocks of black and white ducks of two or three kinds, wood ibises, blackish ones, like those that we shot at Barranquilla, roseate spoonbills, white egrets, snowy herons, cocoi herons, the black and white terns, some small sand snipe, large plover, three kinds of kingfishers, numbers of the screamers, caracara eagles, etc. Late in the afternoon we stopped for wood, and Cabell and myself hurried ashore, but in a few minutes a drenching rain fell, and before we could run back to the boat we were soaked. Cabell shot a very large hawk, probably a young caracara eagle, as it had the same large bluish white beak with pinkish colorations along the side of the head. It was in wretched plumage, and stunk so intolerably of carrion that we did not bring it on board. It was of a dirty dark brownish color above, with a great many narrow brown and white bars on the tail. In the same tree in which the hawk was sitting there were three large iguanas. They seem very abundant along the river here.

The boat ran all night. Just about dusk, as we were running close to the shore, a large yellowish owl flopped out from some scrubby bushes and flew off from the river. With the exception of the rain in the afternoon, it was clear and hot.

Wednesday, July 20, 1892. When we woke at daybreak this morning, our boat was just making fast to the wharf at Barranquilla. We dressed quickly, got off our baggage, and drove around to Miss Hoare's, where we were given very nice rooms. After we had taken some coffee, Alice rested whilst Cabell and I walked around to the market. I bought two more jaguar-skins, — not such large ones as the one I had gotten at Honda, — and paid for them seven dollars and fifty cents in paper. I saw in the market an Indian with a macaw of a kind that I had not seen so far. It was large, with scarlet and yellow the prevailing colors, the wings being largely yellow (*Ara macao*). I was told that it had been caught near Barranquilla.

Later on it grew so hot that we returned to the hotel, and sat around in the shade until about four o'clock, when we went out for

CATHEDRAL AT BARRANQUILLA.

a long drive through the town. It covers a good deal of space, but except the cathedral and one or two buildings near by, there are no houses of any architectural pretensions. The majority of the dwellinghouses are of mud and bamboo, thatched with rushes. In some the mud is whitewashed, or is plastered smooth with lime, so that from the street they appear solidly built of brick. One house of some pretension was unique. Between each window was a large panel in the wall, and these panels were decorated with paintings or frescoes horribly executed, the figures of life-size and gaudily colored. In one of the panels was the "Angelus." In the evening we called on Captain Bradford. It was clear and hot.

In the courtyard of our hotel there was a little armadillo which had been bought at the market here. Its body was about the size

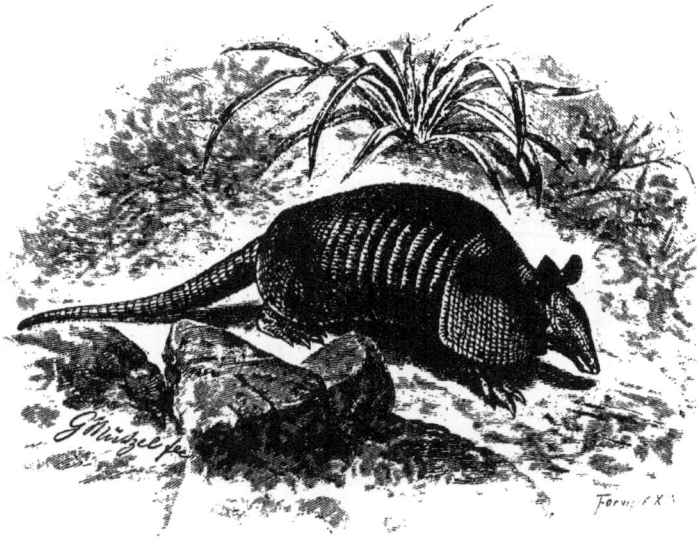

NINE-BANDED ARMADILLO (TATUSIA NOVEMCINCTA).

of a rabbit's, but its head was longer and smaller, and its tail long and thick at the base. Its scales did not overlap, nor were they separate, but were like those on the under parts of an alligator. Its eyes were very small (*Tatusia novemcincta*).

Thursday, July 21, 1892. Cabell and I had arranged for a man to come and take us out in a boat early this morning, so we might get some waterfowl; but although we were ready by five o'clock, he failed to appear. About seven we walked around to the market, where, after considerable inquiry, we managed to find a man and a boy, who took us out in a large dug-out. We started about half past seven, and went through a large ditch or canal, which ran across a marshy tract for over a mile, until it opened into the Magdalena. Here we turned up-stream, and hugged the shore closely so that our boat could be poled along. On our way in the canal we came across the badly decomposed body of a very large alligator, in whose jaws a few teeth yet remained, which our boat-man pulled out for us. As we turned a bend in the canal, we saw coming towards us a boat-load of natives transporting cattle in a most curious way. The boat was a huge dug-out, but was so narrow that four or five bullocks would have filled it, so the boatmen had devised a peculiar plan. They had lashed across the boat, at equal distances apart, three long poles that projected like out-riggers ten feet or more on either side. These poles were probably about a foot above the surface of the water. The cattle were driven into the water until they were swimming, and then their horns were lashed firmly to the poles. For each pole there were eight bullocks, — four on a side, making twenty-four in all. The boat was poled along by the crew, the cattle swimming, and the poles keeping their heads above water, so that they could not drown.

We had hardly left the canal when I shot, on a mud flat, a small grayish heron. It was smaller than our green heron, but quite similar in coloration, the top of its head dark; its back and wings the same greenish gray, with lighter edgings to the feathers; the neck light and streaked below (*Butorides cyanurus*). This bird was unfortunately stolen by a cat at the hotel before I had skinned it. At this same spot I saw standing, on a strip of mud by a pool in the marsh, one of the white-winged jaçanas that are so common here. I shot and killed it, and then directed the little Indian with us to bring it. He started for it; but before he could reach it a small alligator darted out of the pool, grabbed the bird, and returned to the water with it. I ran up at this, and frightened the " caymancito " so that he dropped the jaçana, but rose to the surface, and, with his eyes just on a level with the water, watched me closely. Cabell now came up with the rifle, and I took good aim at a distance of some ten feet, and blazed away. The alligator turned over on his back, and sunk in about eighteen inches of water. I waded in, and secured both it and the bird. The alligator was stone dead; yet, though I examined it closely a dozen times, I could never detect the slightest scratch upon it. It was probably killed by the concussion, as blood oozed from its throat.

The jaçana had a yellowish orange beak, a scarlet frontal shield and lobes at the side of its mouth. It had a yellow, thorn-like spur on the inner side of each wing at the wrist-joint. Its general color above and below was black, with greenish and purplish reflections; its primaries and secondaries pale, greenish white, with narrow, blackish edgings; its legs and feet olive (*Jaçana nigra*). When I came to skin this bird, I found that, although it much resembles a coot, it is very easily skinned, whilst it is almost an impossibility to get the skin of a coot's neck to pass over the head. Later in the day I saw others that were whitish below. They were probably young. The one that I killed was a female. At this place I saw a slender clay-colored snake; but it ran under some driftwood before I could kill it. A little farther up the river I shot at a purple gallinule, and crippled it, but did not get it. Cabell got a snap shot at a small alligator, but missed, and later he shot one of the terns that we had seen so often. It was a large bird, a male in poor plumage, grayish above; tail short and forked, dark grayish; wings white, primaries black, below white, crown black, beak yellow, feet the color of yellowish green oil paint (*Phaëthusa magnirostris*). Farther up the river, where an opening offered, we went ashore. Here we found among the underbrush a number of small

BLACK JAÇANA (JAÇANA NIGRA).

whippoorwills, and Cabell shot two, both females.   They were much
smaller than ours, had bristles along the gape, their tails were
slightly forked, and marked like that of our night-hawk, except
rusty instead of gray.   Above they were mottled with rufous and
black (*Stenopsis ruficervix*).   Here, also, we got several long shots
at some cormorants, but failed to get any.   They seemed to be of a
uniform grayish color.   The boatman called them " pato cuervo,"
crow-duck.

I saw at least three species of ducks, but could not identify any

MURINE OPOSSUM (DIDELPHYS MURINUS).

of them.   When we returned to the boat we concluded to cross the
river, so we hoisted a coarse sail of bagging, and were soon across,
although the river here is very wide.   As we passed under some

willow-like bushes overhanging the water, I saw in one of them a large mud ball, about the size of a man's head. Cabell pulled it down, and found it to be a nest of some kind. It had a little hole in one side, and was lined with strips of bark. In it he found what I thought at first was a rat; but a glance at the thumb on the fore paws showed it to be a 'possum. It was smaller than a rat, yellowish brown above, paler beneath, with a black stripe on each side of its head, from its nose to its ear, embracing the eye (*Didelphys murinus*). It was a female, I think, and had no pouch that I could discover; but its teats were arranged in a circle at the lower part of its belly. At this place we saw a number of large alligators, but got no shots at them, as they were all swimming. We went ashore here, and I shot one of the smallest-sized kingfishers (*Ceryle americana*). It was just a miniature of the one that we got on our way up the river, glossy green above, white below, a white collar, and a chestnut-red belt. This was a male. The female which I saw was without the belt. We saw numbers of iguanas and of the large brown hooded lizards. The latter were called "lobos," or wolves, by our boatmen.

It had now become too hot to remain out longer, so we turned back, and reached the market about eleven. On our way we passed several dug-outs loaded almost to the water's edge with mangoes, which are eaten here in great quantities. We saw birds too numerous to mention : great flocks of parrakeets, numbers of kites, herons, ducks, etc.

In the market we found some men skinning a manatí, which they had just harpooned in the river. It was upon its back, and was half skinned, so I could not get a good look at it. It was about seven feet long and as thick through as a small horse. Its color was that of a hippopotamus, its skin very thick with a few coarse hairs, its flesh like coarse beef in appearance and covered with heavy blubber. Its tail was flat and fan-shaped, with no divisions, and not so pointed as in the figure given. Its fore flippers were like long paddles and smooth, but when skinned, the different bones were

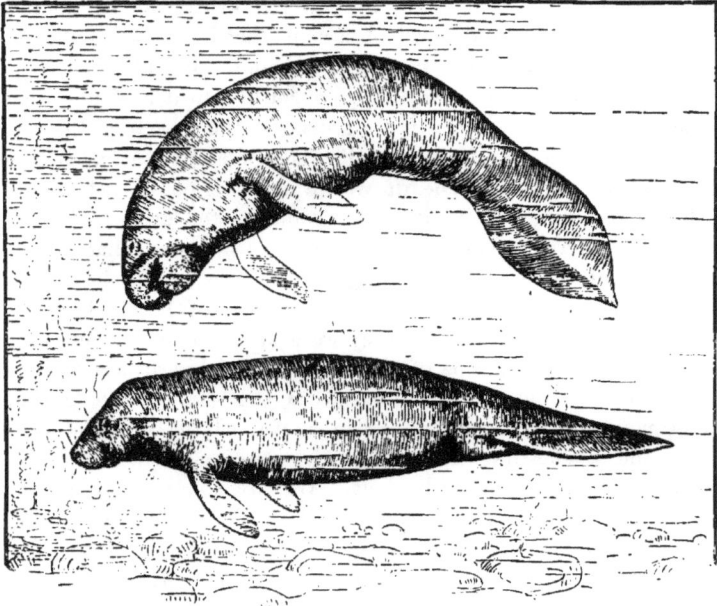

THE MANATÍ.
(From U. S. Fish Commission Report.)

easily seen. Its lips were thick and fleshy, and well covered with
stiff hairs.   I promised the man a reward if he would bring me the
skull, but he failed to do so.   When we reached the hotel we had
breakfast, after which I skinned the birds and the 'possum.

We had intended to leave for La Guayra by the French steamer,
and take the Red " D " Line from that place on the 30th, but fortunately for us, the Royal Mail Steamer, the Derwent, was in port,
and would leave for Curaçao on the next day, so we concluded to go
there instead, especially as we heard bad accounts of the state of
affairs in Venezuela due to the revolution.

Late in the afternoon we went for a drive, and upon our return

Alice began to feel badly, and spent a sleepless night, troubled with a high fever and severe headache. It was clear and very hot.

Friday, July 22, 1892. Just as soon as possible after daybreak, I secured a doctor, who found Alice suffering from malaria, and prescribed quinine. Cabell and I spent the morning in packing, and at one o'clock left our hotel for the station. Our train left shortly before two, and we reached Puerto Colombia in about an hour. We noticed a great change in the country since we first went over the road. Everything now was dry and dusty, where there were pools and lakes before. We saw at one place a lot of large plover, and some stilts that waded about in water up to their bodies. The ride down was extremely hot, and was very trying to Alice, who had a slight chill when we reached the wharf. We had to wait over an hour for the tug, which finally came and carried us out to the Derwent. Captain Buckler gave us very good staterooms, and I got the ship's doctor to prescribe for Alice as soon as possible.

The steamer started about five, and in a few minutes I had to give in to seasickness. Cabell, however, kept well.

It was clear and hot.

CHAPTER VII.

CURAÇAO AGAIN.

SATURDAY, July 23, 1892. This morning at daybreak we had a magnificent view of the Sierra Nevada of Santa Marta. I could not believe at first that the immense snowy masses that apparently towered above us could be anything else but clouds. They appeared to rise abruptly from the seashore, but in reality they are some thirty miles inland. The highest peak is over 16,000 feet above the level of the sea. As the sun rose, clouds began to gather, and soon blotted out from our sight the last portions of the Colombian coast. Alice was feeling better at night, though still suffering from headache. I was seasick all day, so sat around and confined myself to a diet of toast and ginger ale. There was a perfect menagerie on board, belonging to the officers and crew : a tiger cat, a peccary, two monkeys, a red and white squirrel, a pair of thick-billed euphonias, and about twenty macaws, parrots, and parrakeets. The tiger cat, which was of the same species as those of which we purchased the skins on the Magdalena, was taller and slightly larger than our wild-cat, and very prettily marked.

It was clear and hot.

Sunday, July 24, 1892. Early this morning we passed on our right the island of Aruba. It is similar in appearance to Curaçao, which place we sighted about noon. We finally entered the harbor of Santa Ana between four and five. I went ashore as soon as possible, and secured some good rooms at the Hotel Commercio ; then returned, got together our baggage, and we left the ship. We had a large, bright room through which the trade wind blew steadily, and we found it delightfully cool, and enjoyed a good night's rest.

It was clear and hot.

Monday, July 25, 1892. We stayed in or near the hotel the greater part of the day, and did nothing in particular. We secured staterooms on the Venezuela, which was expected on the 28th. After breakfast Cabell and I took a short walk. A negro fishing on the wharf near the old fort gave me a very beautiful fish. It was about the size and somewhat of the shape of a large sun perch, black, with bright yellow stripes, and a brilliant blue line just below the eye. Each of the black scales had a little silvery crescent at the tip (gen. *Pomacanthus*).

About four o'clock Alice and I went for a short drive to the southeast of the town. We found splendid roads and everything looking fresher and greener than when we were here last. The houses and streets were certainly very clean and prettily kept. I noticed at one place a gang of convicts sweeping the streets. They wore the ball and chain, but were not dressed in the striped clothes that our convicts sometimes wear. Instead, they wore a white coat with one red sleeve and one blue sleeve, and the legs of their trousers were of different colors.

Houses, even those in the outskirts and far into the country, are solidly built as a general rule, but we saw a few little wattle and thatch huts which looked quite picturesque.

I saw flying over the harbor a flock of five brown pelicans, and some medium-sized terns, gray, both above and below, with black crowns.

It was clear and warm.

Tuesday, July 26, 1892. I woke about half past five, called Cabell, and we took our guns, crossed the harbor, and went up over the hill to the spot where we hunted when here before. We saw the same species of birds that we had seen, and, in addition, some others. I saw quantities of the little ground doves, the little dark finches, the honey-creepers, and the chestnut-crowned yellow warblers, of which I killed three, two males and a female. The female had no chestnut on the head, and the spots on the breast were very faint, almost wanting. I also saw quantities of the chestnut-collared sparrows, and shot a male. As we were walking along the road near the monastery, I saw a pair of partridges run through the hedge in front of me, and by a snap shot I killed one, a male in fine plumage. It was very similar to those that we had gotten in Guaduas, perhaps smaller, and lighter colored generally, its throat, chin, and forehead being buff without the rufous of the Guaduas bird (*Eupsychortyx cristatus*). Cabell shot one of the mocking-birds, which I found to be very similar to ours. This was a young bird with speckled breast (*Mimus gilvus rostratus*).

Upon one of the hills we came across a flock of seven or eight large pigeons, the "ala blanca." They seemed to be large grayish

WATTLE HUT, CURAÇAO.

birds, with a white streak in each wing. Cabell got a long shot and struck one, but failed to get it. They were very shy, and we got no more shots at them. We also saw a number of doves or pigeons of a smaller size, intermediate between this and the ground dove. They were all flying at a distance and we got no shots. Later I shot a sparrow-hawk, a female, marked like ours, above reddish brown with black bars, its head bluish gray above with traces of rusty on the scalp ; below whitish, streaked with brown, its thighs whitish, cere and feet yellow, eyes brown, beak horn-blue, black at the tip (*Tinnunculus sparverius brevipennis*).

When we reached the tamarind-trees near the old convent, we found them in bloom and perfectly swarming with humming-birds. In a few minutes we got twenty-four ; but as there were but two species among them, we killed no more. They were the ruby and topaz (*Chrysolampis moschitus*) and the small emerald green (*Chlorostilbon atala*). We started back shortly after eight, and on the way we flushed some more partridges, but failed to get any. We found in a scrubby cactus the nest of one of the little dark-colored finches. It was almost spherical, with a hole in one side, and contained three eggs much like those of our field-sparrow. The ruby and topaz humming-birds that we killed were in very poor plumage, their bodies covered with undeveloped pin-feathers. We reached the hotel about half past nine, and after breakfast worked for several hours skinning our birds. In the afternoon we went for a long drive, this time in a northerly direction and beyond the monastery. Everything looked beautiful. The roads were well kept and ran between high hedges of the club cactus. Notwithstanding the mountains, there is a good deal of level land. In all of the little valleys there are small embankments, or dams, built across at intervals, — just as we make ice-ponds, — to catch any rain that may fall, and let the water soak in instead of running off. I saw quantities of the lizards that I saw when here before. It was cloudy in the morning, and rained after breakfast, but held up before we went out for our drive, so we found it very cool and pleasant.

There are three or four papers published in Curaçao, and one that I came across was printed in the Papamiento dialect. After I had read some of it, I did not wonder that I found it difficult to under-

MOUNTAIN AT CURAÇAO.

stand. I give below an advertisement with the corresponding Spanish, so that they may be compared : —

Plateria . . .	Platería.
Caya Grandi.	Calle Grande.
E winkel aki ta ofrecé na pú-blico su sirbishi, garantizando tur trabauw pa bon ejecucion i bon gusto. També tin di beende un gran surtido di, etc.	El —— aqui á ofrecer al pú-blico sus servicios, guarantizando todo trabajo para buen ejecución y buen gusto. Tambien tiene y viende un gran surtido de, etc.

In Papamiento the name Curaçao is spelled Corsouw. It is a

phonetic derivation from Spanish with a mixture of Dutch (as in the word winkel above), but in many cases there is an omission of syllables, as, for instance, " tur lo ke ta " is in Spanish " todo lo que está." Whilst I am on the subject of languages, I am reminded to say that in Colombia a good Spanish is generally spoken. There are a few peculiarities that I noticed. The letter *c* is generally pronounced as in English without the lisping *th* sound that I had been taught was proper in Spain. The *d* in such words as lado, pescado, colorado, etc., is generally omitted in pronunciation, thus making the word lao, pescao, etc. The letters *b* and *v* are interchanged in a hopelessly confusing way ; *b* is in general pronounced *v*, but the rule to the contrary is sometimes observed, as, for instance, the word for twenty, veinte, I heard pronounced beinte. There are some delicate shades of meaning expressed by uses of the augmentative and diminutive terminations. Temprano means early, tempranito means very early, or soon in the morning : " ¿ Está todo arreglado ? " means, Is all arranged, or ready ? " ¿ Está todito arreglado ? " Is every single thing ready ? The ordinary appellation for servants is hombre, in preference to mozo.

Wednesday, July 27, 1892. Cabell and myself went out early this morning to the same place that we visited the day before. When we reached the spot where we had seen the pigeons, I crept up cautiously, got a long flying shot and killed one, a male. It was a large bird, the size of a common pigeon, its beak light flesh-color, the nostrils pinkish, eyes reddish, skin around them blue, around this a circle of brown roughened skin looking like the sandpaper on a match-box. It was of the usual dove-color, becoming bluish on the rump, lighter below, the tail plain grayish with no bars or marks, the scapulars brownish gray, a diagonal white band from the wrist-joint to the scapulars, the primaries and secondaries sepia with fine white edges. The feathers of its iris were prettily marked ; each was something like a miniature turkey's feather ; a narrow band of black at the tip, and above this a strip of metallic color, giving the neck a barred appearance. The feet were large and of a deep pinkish red (*Columba gymnopthalma*). A little farther on Cabell shot an oriole similar to the one we had killed at Barranquilla. It was the size of our Baltimore oriole, brilliant yellow, its throat, chin, spot from eye to beak, tail, and wings black, wing with a white bar and many feathers edged with white, some of the tail-feathers white tipped (*Icterus xanthornus curasoënsis,* Ridgw.). This was the bird that our guide had called a troupial when we were here before. A little later Cabell shot another, which was also a male, but was dull olive-yellow, darker on the tail and wings and lighter below. We also got another mocking-bird and a honey-creeper like the one that we killed before, but having its throat and supra-orbital stripe yellow.

From this place we crossed over to the seashore, where, on the edge of some salt-pans, I shot a young male spotted sandpiper (*Actitis macularia*). This was in the unspotted plumage, its breast plain white. We saw very many humming-birds, but shot none, as they were all of the same species as those that we had killed the day before. We also saw in a mangrove swamp a pair of small herons,· which I took to be our green heron, a number of the medium-sized doves, a medium-sized tern, apparently pure white, and a very small one, white with a black crown. We reached the hotel before breakfast, and afterwards skinned our birds. In the afternoon we went for a long drive. I carried my camera and took a number of views. We saw great numbers of humming-birds and ground doves, and three partridges running along the road with their crests up. I also saw several sparrow-hawks, and a hawk of much larger size flying at a distance.

Donkeys are used a great deal around Curaçao. We saw many men and women riding them, the women sitting astride of the little animals, with their big toes thrust in loops of cord which served as stirrups. It was partially cloudy and cool.

Thursday, July 28, 1892. The Venezuela came in shortly after daybreak this morning, so as soon as we were up we went over to see our friends on board. After our return to the hotel I took my

J G Keulemans del.   Mintern Bros Chromo lith London

ICTERUS XANTHORNUS CURASOENSIS. *Ridgw.*

Curacao Oriole.

camera and went up to Fort Nassau on the hill back of the town, whence I took several views of the harbor. I passed on my way some tamarind-trees in bloom, around which were quantities of humming-birds of the two species. I also saw numbers of the chestnut-crowned yellow warblers, the honey-creepers, and the

ABORIGINES OF CURAÇAO.

mocking-birds, and three sparrow-hawks. In the afternoon we went for a long drive to the southeast of the town. At a point on the southern edge of the Lagoon we passed some large shallow salt-water pools, where I saw wading about sand snipe of three different sizes, but could recognize none of them. I also saw at a distance a large hawk that flew like our marsh-hawk. I regret that we could not stay here long enough to work up thoroughly the birds of the island. In the course of our drive we passed many attractive-looking country places surrounded by cool groves of fruit and palm trees. The nísperos were in perfection. They much resemble a russet apple, but are soft and pulpy, with large flat seeds. The pulp is very sweet, like unrefined sugar, and though I did not care for them at first, I soon grew to like them. I also tried some cashews, or cachús, and found them not unpleasant, but my unfortunate curiosity led me to bite into one of the kidney-shaped excrescences at the larger end. Be warned by me, and if you ever have a cachú, avoid the bean.

DONKEY TEAM. CURAÇAO.

Of all the most disgusting, acrid, bitter, burning, clinging tastes this is the worst, and though I went no farther than to stick my teeth into it, it was hours before I could rid myself of the taste, even though I repeatedly rinsed my mouth with pure alcohol.

During our stay here we saw many fish in the harbor, but they did not seem to bite at all. Great numbers of a fish not much larger than a sardine are caught in cast nets. Some of them are dried and sold thus, tied up in little bundles like cigars (*Trachurops sp.*). Very good food fish are caught outside. Among them I noticed "el capitan," a fish much like our scup, about a pound and a half in weight, and with red pectoral and ventral fins, and the "king fish," eight or nine pounds in weight, a species of mackerel of a uniform dark color. Of the shellfish, I saw quantities of sand fiddlers, a few crabs of larger size, deep mahogany red with white claws, and some large lobsters, much like ours but without the two big claws and with very long antennae. There are said to be some rabbits on the island and a few snakes, but we saw none.

The natives here have a peculiar way of hitching two donkeys to a cart. One is put between the shafts, whilst the other has no other harness than a loop around its neck, one end of which is tied to the nearest shaft. Oxen are not yoked as with us, but a cross-bar is lashed to their horns, the weight thus coming just on their foreheads. It was clear and hot.

Friday, July 29, 1892. As the Venezuela was to leave in the afternoon, we were busy all the morning, cleaning our guns and giving a final packing to our baggage. We finally went aboard about two o'clock, but it was not until after six that we left the harbor and headed for La Guayra. We had supper immediately after getting outside, and when we came up on deck afterwards it had grown dark, and the island of Curaçao had faded from our view. It was clear and hot.

Saturday, July 30, 1892. When I went out on deck this morning we were within a few miles of La Guayra. The view was beautiful. The town lay on a narrow strip of land at the foot of a mountain that rose abruptly from the sea until its top was hidden from us by clouds. A few houses above the town were actually built in niches which had been excavated to receive them. To our right we could trace by the cuttings the railroad winding its way up to Caracas; to our left lay the main portion of the town, above which

BREAKWATER AND HARBOR OF LA GUAYRA.

was seen the roof of the bull-fighting arena, and higher up on the mountain-side a little pill-box of a fort. There was a Venezuelan man-of-war in the harbor, a dirty little steamer about the size of the average steam yacht. We tied up alongside a strong-looking pier and breakwater of concrete and iron, out upon which ran the tracks of the railroad, a narrow-gauge road with English cars and locomotives, the passenger coaches looking like a second-hand summer street car. Shortly after we had tied up, Cabell and I took a short walk up into the town. We found it indescribably filthy and bad smelling, the stores dirty, narrow, dark, and overhung with cobwebs. There was an air of general stagnation of business, due no doubt to the revolution then in progress. There is a fine stream tumbling down the mountain through the centre of the town. It is walled in on either side. We found the heat so oppressive that we soon returned to the ship. On the wharf we were much interested in seeing the fishermen come in. They go out to sea in little cockleshell dug-outs of a different design from those used at Barranquilla. These are skiff-shaped, ride very high in the water with both ends clear, and are painted. The fishermen squat on the bottom in the middle of the boat, and, using a single-bladed paddle which they change from side to side about every third stroke, they skim swiftly over the water. They brought in some fine fish, some that I recognized, others that I did not. Among them were several fine red snappers and Spanish mackerel; some fish of the mackerel species, about a yard long, with heavy teeth and of a uniform dark color (*Cero sp.*); some perch-like fish with yellow longitudinal stripes (*Pomacentrus sp.*); a small brown fish very like our chogset, but with circular dots of sky-blue all over its body (*Hæmulon sp.*); a few small flat fish, and an eel, broad and thin, brown with light yellow dots, a wide opening mouth with vicious-looking teeth (gen. *Muræna*). The water was marvelously clear, and looking over from the pier we saw some of the most beautiful fish that I have ever seen. There were some little fish marked with broad black and yellow bars like a sheepshead, some fool-fish (*Alutera sp.*), and

some of the brilliant yellow and black ones like the one that I saw
in Curaçao; but the most beautiful of all was what the men on the
wharf called "loro," or parrot. As well as I could see, it was of
the same shape as the black and yellow ones, its head and neck a
vivid blue, its body light green, its tail a golden yellow, and its fins
tipped with pink (family *Scaridæ*). We had some flying-fish for

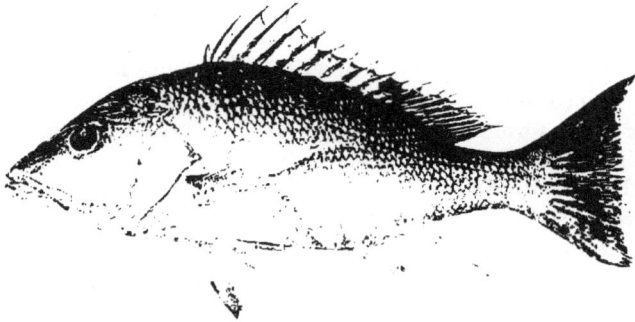

RED SNAPPER.

breakfast on the steamer. They taste a little like smelts, but are
dry. Whilst we were watching the fish a frigate pelican hovered
over our heads for some time, coming at times within thirty yards
of us so that we had a good view of it. The small boys on the
wharf were catching great numbers of the little fish that were so
abundant at Curaçao (*Trachurops*). We left La Guayra about
eleven o'clock, and headed north. The vessel ran steadily, and I
fortunately escaped seasickness. Before sunset we passed Los
Roques.

The rest of our trip was uneventful. The sea remained as quiet
as a mill-pond, and we made fine runs of nearly 340 miles per day.
Among the passengers we had the pleasure of meeting our consul
from Curaçao, Captain L. B. Smith, a most agreeable gentleman,
whom I will always regret not having met during our stay on the

island. This gentleman told me that but a few days before he had seen a barn-owl that was caught near the town (probably *Strix flammea bargei*, Hartert). He also told me that on the island of Bonaire there were many flamingoes. It is to his enterprise that the town owes the drawbridge across the harbor, its ice-machine, and other improvements. Since my visit he has sunk five artesian wells upon his place, Planter's Rust, the combined production of which is 87,000 gallons of water of excellent quality per day. This will prove a godsend to the island, as only those who have been there can form an idea of the great importance of this water supply.

We made the Mona Passage on Sunday afternoon. On Tuesday a turnstone flew around us several times and finally went off in the direction of the Bermudas. On the following day a small warbler lit on the upper deck among the life-boats, but I could not find it. On Thursday we saw several schools of blackfish and a whale. At sundown we saw the Barnegat Light, and about eleven o'clock we came to anchor off quarantine. At half past eight on Friday morning, August 5, we landed in Brooklyn, and our trip was at an end.

CHAPTER VIII.

WE had been gone from New York just fifty-four days. In that time we had been sixteen days on the ocean, twelve days on the Magdalena, and three days on the mule road; that is, we were traveling thirty-one days. Exclusive of the shooting that we did during the stops of the river steamer, Cabell and I had been out together with our guns ten times, and he had been out three times alone. We brought back 210 skins. I give below lists of the birds that I observed in Colombia and in Curaçao. Mr. Robert Ridgway of the Smithsonian Institution has been kind enough to identify the skins for me and also to assist me greatly in the preparation of the accompanying lists. The names of those birds identified beyond a doubt are printed in small capitals, whilst those which are at all doubtful are printed in italics. References after notes refer to colored plates of the bird.

BIRDS OBSERVED IN COLOMBIA, SOUTH AMERICA.

1. PHAËTHUSA MAGNIROSTRIS (LICHT.). Large-billed Tern.

Abundant on the Magdalena from Barranquilla to above Puerto Berrío. We often saw as many as a hundred standing together on some of the low sand bars, and sometimes a dozen or more would float by us on a piece of driftwood.

2. ANHINGA ANHINGA (LINN.). Snake Bird.

I saw a dozen or more of these along the Magdalena, but never more than two together. They were usually flying, but a few were seen perched on dead snags. Aud. *B. of N. A.* vol. 6, pl. 420.

3. *Phalacrocorax cigua* (*Vieill.*)? Brazilian Cormorant.

A good many small cormorants were seen flying over the marshes at Barranquilla, but as I did not get a specimen, I am doubtful about the identification. U. S. Astron. Exped. to Chili and Peru.

4. PELECANUS FUSCUS (LINN.). Brown Pelican.

These were the first birds that we saw as we approached our anchorage off the Colombian coast. We saw large flocks of them around the seashore, but none in the interior. The Spanish name is "alcatraz." Aud. *B. of N. A.* vol. 7, ppl. 423, 424.

5. CAIRINA MOSCHATA (LINN.). Muscovy Duck.

These ducks were seen continually from Barranquilla to within a short distance of Yeguas, sometimes in enormous flocks. They often lit in trees when first disturbed by the steamer. The native name is "pato real," royal duck. I also saw ducks of other species, but could not identify them; among them two species of Dendrocygna.

6. AJAJA AJAJA (LINN.). Roseate Spoonbill.

We saw a few small flocks of these lovely birds at different places along the river. The largest contained six individuals. Aud. *B. of N. A.* vol. 6, pl. 362.

7. PHIMOSUS INFUSCATUS (LICHT.). Dusky Ibis.

Mr. Ridgway identifies my description of the two shot at Barranquilla as belonging to this species. I failed to save their skins. We saw a number at Barranquilla, but few higher up the river. I saw some carrying sticks for nests on June 22.

8. TANTALUS LOCULATOR (LINN.). Wood Ibis.

These were seen in great numbers along the river, especially along the central portion. When suddenly disturbed they flew off irregularly in different directions, but when traveling they kept together in strings. They usually flew with heavy wing-beats, but I saw many soaring at a great height with motionless wings. Aud. *B. of N. A.* vol. 6, pl. 361.

9. MYCTERIA AMERICANA (LINN.).   Jabiru.

I saw only two of these large birds, and they were on the lower river. I thought that they were a species of ibis, as they flew in just the same way, with their necks extended instead of doubled back like the herons. Their plumage is snowy white; the beak, head, and legs black. The native name is "cabeza negra," black head.

10. ARDEA COCOI (LINN.).   Cocoi Heron.

Seen abundantly at every point along the Magdalena. This bird is very much like our great blue heron, but has more white below, and the entire crown is black. The Spanish name for heron is "garza."

11. ARDEA EGRETTA (GMEL.).   White Egret.

I saw a great many of these birds along the river, though they were by no means as abundant as the preceding species. We sometimes saw them in small flocks, but rarely saw more than two of the cocoi herons flying together. Aud. *B. of N. A.* vol. 6, pl. 370.

12. ARDEA CANDIDISSIMA (GMEL.).   Snowy Heron.

These beautiful little birds were by far the commonest of the herons. We saw them continually, and at some places the muddy edges of the river were lined with them. Towards nightfall they flew overhead, going to their roosting-places in large flocks. Aud. *B. of N. A.* vol. 6, pl. 374.

13. BUTORIDES CYANURUS (VIEILL.).   Blue-tailed Heron.

I saw a good many of these in the marshes near Barranquilla, and I shot one; but unfortunately it was stolen by a cat before I had skinned it. They are smaller than our green heron, but similar.

14. IONORNIS MARTINICA (LINN.).   Purple Gallinule.

I saw a few in the marshes at Barranquilla, and wounded one, but failed to get it. Aud. *B. of N. A.* vol. 5, pl. 303.

15. HIMANTOPUS MEXICANUS (MÜLL.).   Black-necked Stilt.

From the train I saw several pairs of these birds wading in

some pools a few miles below Barranquilla. I saw them both in June and in the latter part of July. Aud. *B. of N. A.* vol. 6, pl. 354.

16. JAÇANA NIGRA (GMEL.). Black Jaçana.

These birds were abundant in the marshes around Barranquilla, especially where there were lily pads floating on the surface of the water. They were very noisy, and often held their wings up vertically, as some snipe do, as if stretching. I saw some with their under parts lighter colored, probably young.

17. EUPSYCHORTYX LEUCOTIS (GOULD). White-eared Partridge.

The two killed by my brother at Guaduas were identified by Mr. Ridgway as belonging to this species. We saw numbers of partridges at Guaduas, at Honda, and at Barranquilla, but did not succeed in getting others, so cannot tell if they were all of the same species or not. We found it impossible to flush them a second time; and it so happened that whenever we got shots, our guns were loaded with dust shot, so we failed to stop the birds. At Barranquilla I heard partridges uttering the familiar call "bob-white." Gould, *Mon. of Odontophorinæ.*

18. *Stegnolæma montagnii (Bonap.)*? "Guacharaca."

I saw one of these domesticated at Honda, and lower on the river I saw a small flock in the edge of the forest. The identification is from my meagre description, and therefore is very doubtful.

19. CHAUNA DERBIANA (GRAY). Colombian Screamer.

We saw a few of these birds on the lower Magdalena. They were either perched in the tops of dead trees, or walking about on the ground like turkeys. I saw a pair domesticated. They kept with the poultry, and walked about in a very slow and dignified manner. Pl. 11, P. Z. S. 1864.

20. COLUMBA RUFINA (TEMM.). Green-naped Pigeon.

I shot a fine specimen on the Magdalena, and my brother killed a second one at Guaduas. These were the only ones that I saw.

21. ZENAIDA RUFICAUDA (BONAP.).   Rufous-tailed Dove.

We killed a couple of these doves at Honda, both of which were males; and we saw them frequently during our stay at that place.

22. COLUMBIGALLINA PASSERINA (LINN.).   Ground Dove.

We found this little dove common at Barranquilla, Honda, and Guaduas. When running along on the roads, they carry their tails held up very prettily. Aud. *B. of N. A.* vol. 5, pl. 283.

23. COLUMBIGALLINA RUFIPENNIS (BONAP.).   Rufous Ground Dove.

We saw a few of these at Barranquilla and at Honda. They are not so abundant as the preceding species.

24. GYPAGUS PAPA (LINN.).   King Vulture.

We saw three individuals on the Magdalena a short distance above Barranquilla. They were all perched in dead trees, which grew in overflowed marshes. Descourtilz, *Orn. Brésilienne.*

25. CATHARTES AURA (LINN.).   Turkey-buzzard.

Common at Barranquilla, Honda, and Guaduas, but not so abundant as the following species. Aud. *B. of N. A.* vol. 1, pl. 2.

26. CATHARISTA ATRATA (BARTR.).   Black Vulture.

Very abundant at every point that we visited in Colombia. They collect in immense numbers around slaughter-houses, and on sand bars in the river when they observe a fisherman cleaning his catch. Aud. *B. of N. A.* vol. 1, pl. 3.

27. ROSTRHAMUS SOCIABILIS (VIEILL.).   Everglade Kite.

These were very abundant at Barranquilla, and flew about over the marshes just as do our marsh-hawks. Baird, Cassin & Lawrence, *B. of N. A.* pl. 65.

28. RUPORNIS MAGNIROSTRIS (GMEL.).

My brother killed one at Guaduas, which was the only one that I saw. The natives called it a " garrapatero," or tick-eater; but they apply this name to the milvago and also to the ani.

29. *Heterospizias meridionalis (Lath.)* ?   Rufous Buzzard.

Mr. Ridgway identifies thus the large rufous-winged hawks that I saw over the grassy meadows at Yeguas. Having nothing but my description to go by, I have indicated the identification as doubtful.

30. *Falco sparverius (Linn.)* ?   Sparrow-hawk.

The remarks for the preceding species apply to this. Those that I saw from the train above Yeguas were near enough to distinguish the crescent marks on the head, and to all appearances were the same as our species. Aud. *B. of N. A.* vol. 1, pl. 22.

31. *Polyborus cheriway (Jacq.)* ?   Audubon's Caracara.

This large carrion hawk I saw at a number of places along the river, and on our down trip my brother shot a young one in poor plumage; but it stunk so from its last meal that I did not skin it. Aud. *B. of N. A.* vol. 1, pl. 4.

32. MILVAGO CHIMACHIMA (VIEILL.).   "Chimachima."

This carrion hawk was common around Barranquilla and at other points higher up the river. They were noisy, and, whilst uttering their cries, held their heads back until it seemed that they would topple over backwards.

33. PANDION HALLÆTUS CAROLINENSIS (GMEL.).   Fish-hawk.

I saw a few fish-hawks along the upper Magdalena. Aud. *B. of N. A.* vol. 1, pl. 15.

34. BUBO MEXICANUS (GMEL.).   Striped Horned Owl.

My brother shot one that was roosting in some thick coffee plants at Guaduas.

35. ARA ARARAUNA (LINN.).   Blue and Yellow Macaw.

This was by far the commonest macaw seen, and was abundant as far up the Magdalena as Yeguas, where the heavy forest ended. Their discordant cries woke us in the mornings, and we saw many of them flying to roost just before sunset. I saw a partly fledged one at Mompos on June 25. Descourtilz, *Orn. Brésil.*

36. ARA MACAO (LINN.).   Red and Yellow Macaw.

I saw in the market at Barranquilla an Indian with one of these macaws, and was told that it had been caught a short distance up the river.

37. *Ara chloroptera* (*Gray*)?   Blue and Red Macaw.

I saw frequently along the river a large macaw, blue, green, and scarlet, but without yellow on the wings.   It may possibly be of this species.   The general name for macaw is "Guacamayo."

38. ARA SEVERA (LINN.).   Severe Macaw.

I saw but the two specimens which I shot on June 28. Descourtilz, *Orn. Brésil.*

39. CONURUS ÆRUGINOSUS (LINN.).   Gray-faced Parrakeet.

This parrakeet was extremely abundant around Barranquilla; but I did not see it higher up the river.   Flocks flew over the town in a steady stream about daybreak, and just before sunset.

40. BROTOGERYS JUGULARIS (DEVILLE).   Orange-chinned Parrakeet.

I saw large flocks of this parrakeet all along the Magdalena as high up as Honda.

41. PSITTACULA CONSPICILLATA (LAFR.).   Blue-rumped Parrakeet.

This little parrakeet I saw along the upper portion of the Magdalena, in some cases associated with flocks of the preceding species.   They were common at Guaduas.   They fly just like English sparrows.   Pl. in this work.

42. PIONUS MENSTRUUS (LINN.).   Blue-headed Parrot.

I saw this parrot only once, when Mr. Lindauer shot one out of a small flock.   This was not far below Yeguas.

43. *Amazona panamensis* (*Cab.*)?   Common Green Parrot.

I saw everywhere at Barranquilla, Honda, and Guaduas in the huts of the natives a green parrot with yellow forehead and scarlet wing edgings.   It was probably of this species, though, as I obtained no specimen, I have marked it doubtful.

44. *Crotophaga sulcirostris* (*Swains.*)?   Grooved-bill Ani.

I saw these birds in abundance at Barranquilla, Honda, and Guaduas; but though I shot several, they were all in poor plumage, so I brought back no specimens, and am now doubtful whether they were of this species or *C. ani*. I saw them feeding on the swarms of grasshoppers at Honda, and I had two of their eggs given to me on June 28.

45. RAMPHASTOS CITREOLÆMUS (GOULD). Citron-breasted Toucan.
The first one that I saw was killed by Mr. Lindauer on June 30. Along the river near this place I saw several others. The peacock-blue color of their eyes is peculiar. I saw at La Guayra a species of cacique with similarly colored eyes. Gould, *Mon. of Ramphastidæ*.

46. RAMPHASTOS AMBIGUUS (SWAINS.). Green-faced Toucan.
I saw a large flock of these toucans at Puerto Berrío, but our steamer stopped there such a short time that I killed only one. Gould, *Mon. of Ramphastidæ*.

47. PTEROGLOSSUS TORQUATUS (GMEL.). Collared Araçari.
I shot two of these and saw five or six others on the Magdalena a short distance below Yeguas. Gould, *Mon. of Ramphastidæ*.

48. BUCCO RUFICOLLIS (WAGL.). Rufous-throated Puff Bird.
I saw a good many of these near Barranquilla, and a few higher up the river. They sit quietly on a dead twig, and look much like small kingfishers. Sclater, *Mon. of Jacamars and Puff Birds*, pl. 29.

49. BUCCO SUBTECTUS (SCL.). Narrow-banded Puff Bird.
I saw but the one specimen that my brother shot on June 28 on the lower Magdalena. Sclater's *Monograph*, pl. 27.

50. GALBULA RUFICAUDA (CUV.). Rufous-tailed Jacamar.
I saw about a half dozen of these birds on the Magdalena. They sit about quietly like kingfishers. My brother saw at Consuelo a jacamar which he described as larger and brilliantly colored, but we did not get a specimen. This was probably *Jacamarops grandis*. Sclater's *Monograph*, pl. 4.

51. CERYLE TORQUATA (LINN.). Great Rufous-bellied Kingfisher.
52. CERYLE AMAZONA (LATH.). Amazonian Green Kingfisher.

53. CERYLE AMERICANA (GMEL.).   Brazilian Green Kingfisher.

These three kingfishers we found abundant from the mouth of the Magdalena until we left the river at Honda.   The last was not so common as the first two; but we found it at Guaduas, where we did not see the others.   Sharpe, *Mon. of the Alcedinidæ*, vol. 1.

54. CENTURUS TERRICOLOR (V. BERL.).   Berlepsch's Woodpecker.

My brother shot a specimen at Guaduas, where I also saw several.   I saw one enter a hole in a dead tree, so it was probably nesting.   At Barranquilla and at Honda I saw various woodpeckers, but did not obtain specimens.

55. STENOPSIS RUFICERVIX (SCL.).   Rufous-necked Goat-sucker.

We found a small flock of these among some stunted bushes near Barranquilla, and obtained two females.   Pl. 14, P. Z. S. 1866.

56. NYCTIDROMUS ALBICOLLIS (GMEL.).   "Parauque."

I saw but the one which I shot on the Magdalena on June 28. At night, along the river, we often heard the cries of various night-birds, some of them very like our "whip-poor-will."

57. NYCTIBIUS GRANDIS (GMEL.).   Grand Potoo.

I saw but the one which my brother shot below Yeguas on July 16.

58. GLAUCIS HIRSUTA (GMEL.).

I saw a good many of these humming-birds along the river. They were in the heavy forests, and fed on the blossoms of a species of canna which grew in the glades near the water.   A female that I shot on June 28 had white feathers scattered about among the green of the back.   On the same day I found one of their nests, but it did not contain eggs.   It was woven to the swinging tip of a plantain leaf.   Humming-birds.   Gould's *Monograph*.

59. PHÆTHORNIS SUPERCILIOSUS (LINN.).

I saw two or three in Guaduas, and shot one which was in poor plumage.

60. LAMPORNIS NIGRICOLLIS (VIEILL.). Black-throated Humming-
  bird.
  We found this bird common in Guaduas.
61. HYPUROPTILA BUFFONI (LESS.). Buffon's Humming-bird.
  We obtained four or five specimens at Guaduas.
62. ACESTRURA MULSANTI (BOURC.). Mulsant's Humming-bird.
63. ACESTRURA HELIODORI (BOURC.). Heliodore's Humming-bird.
  These two species were about equally common at Guaduas.
  From their small size they were very difficult to find when they
  fell in the underbrush.
64. CHRYSOLAMPIS MOSCHITUS (LINN.). Ruby and Topaz Humming-bird.
  We found this species abundant at Guaduas. The full-plumaged
  male was the most beautiful humming-bird that we met.
65. AMAZILIA FUSCICAUDATA (FRASER). Rieffer's Humming-bird.
  I met with this species at two points on the Magdalena and at
  Guaduas, getting four specimens.
66. AMAZILIA CYANIFRONS (BOURC.). Blue-fronted Humming-
  bird.
  My brother shot one at Guaduas.
67. DAMOPHILA JULIA (BOURC.). Julia's Humming-bird.
  We obtained a number of specimens at Guaduas.
68. POLYERATA AMABILIS (GOULD).
  I saw but the one specimen which I shot at Puerto Berrío on
  the river. It had a nest placed on top of a branch in the same
  way that our ruby-throat builds.
69. CYANOPHAIA GOUDOTI (BOURC.). Goudot's Humming-bird.
  I obtained four specimens at one place on the lower river.
70. CHLOROSTILBON ANGUSTIPENNIS (FRASER). Narrow-winged
  Humming-bird.
  I obtained a specimen at Barranquilla, and found it abundant
  at Guaduas.
  Humming-birds were very abundant at Guaduas, but rarely
  until the bird was shot could I tell what was the species. They

could be seen buzzing about in the treetops, but at too great a distance to recognize them, unless they were of peculiar shape or size. My brother saw one at Guaduas which he described as having a scarlet back.

71. MILVULUS TYRANNUS (LINN.). Fork-tailed Flycatcher.

I saw a few at Guaduas and a good many at Honda. Their flight was extremely graceful. Aud. *B. of N. A.* vol. 1, pl. 52.

72. TYRANNUS MELANCHOLICUS (VIEILL.). Melancholy Flycatcher.

73. MYIOZETETES CAYENNENSIS (LINN.). Cayenne Flycatcher.

These flycatchers were common all along from Barranquilla to Honda, and around Guaduas.

74. *Megarhynchus pitangua (Linn.)*? Pitangua Flycatcher.

The large-billed, rufous, and yellow flycatcher which we got at Barranquilla was probably of this species; but as I did not bring back a specimen, I have marked it doubtful.

75. FLUVICOLA PICA (BODD.). Pied Flycatcher.

This conspicuous little bird was abundant in the marshes around Barranquilla, and I saw others at points higher up the river.

76. XANTHOSOMUS ICTEROCEPHALUS (LINN.). Yellow - headed Blackbird.

I saw large flocks of this bird around Barranquilla.

77. ICTERUS ICTERUS (LINN.). Troupial.

I saw troupials in confinement at many places along the Magdalena and at Guaduas, but none in a state of freedom. Aud. *B. of N. A.* vol. 7, pl. 499.

78. ICTERUS XANTHORNUS (GMEL.). Yellow Oriole.

Common at Barranquilla.

79. CASSICUS FLAVICRISSUS (SCL.). Yellow-vented Caçique.

80. OSTINOPS DECUMANUS (PALL.). " Oro péndola."

81. GYMNOSTINOPS GUATIMOZINUS (BONAP.). " Oro péndola."

We got one specimen of each of the foregoing species on the Magdalena on June 28. Higher up the river we saw many straggling flocks of the two last. *Fauna Biologia Centr. Amer.*

82. QUISCALUS ASSIMILIS (SCL.). Colombian Grackle.

This large grackle was abundant around Barranquilla, and often lit in the cocoa palms that grew in the hotel yard. They may have had nests in these palms, but from seeing one with an unfledged young bird in its beak, I am inclined to think that they were robbing the nests of smaller birds.

83. SYCALIS COLUMBIANA (CAB.). Red-fronted Finch.

I saw a few of these near Barranquilla. In our hotel there was one caged which sang very well.

84. VOLATINIA SPLENDENS (BONAP.). Blue-black Finch.

I saw but the one specimen which I shot near Barranquilla.

85. RAMPHOCELUS DIMIDIATUS (LAFR.). Cardinal Tanager.

This tanager was abundant all along the Magdalena and at Guaduas. The native name is "cardinal." *Mag. de Zool.* 1837, pl. 81.

86. RAMPHOCELUS ICTERONOTUS (BONAP.). Yellow-rumped Tanager.

I saw but the one which I shot at Puerto Berrío.

87. TANAGRA CANA (SWAINS.). Blue Tanager.

This tanager is common and I found it from Barranquilla to Honda and at Guaduas. I observed a nest with eggs nearly hatched at Barranquilla in June. The native name is "azulejo," bluebird.

88. TACHYPHONUS MELALEUCUS (SPARRM.). White-shouldered Tanager.

I saw several at Guaduas, and my brother shot one.

89. EUPHONIA CRASSIROSTRIS (SCL.) Thick-billed Euphonia.

I saw but the one specimen which my brother shot on the Magdalena on June 28.

90. PIPRA AURICAPILLA (LICHT.). Gold-headed Manikin.

I saw but the one specimen shot on the Magdalena by Mr. Lindauer on June 29.

91. *Tachycineta albiventris (Bodd.)*? White-winged Swallow.

The little swallow that I saw along the Magdalena may be of

this species. I thought, however, that the body of the bird was white and the wings black, and therefore I leave it doubtful.

The birds which I observed but did not identify would, I think, exceed the above list. Among them were two terns, two ducks, two or three herons, several sand snipe and plover, pigeons, many hawks, an owl, parrots, a toucan, several woodpeckers, humming-birds, many flycatchers, several orioles, finches, warblers, swallows, thrushes, and many wrens.

The following is a list of the humming-birds sent to me from Bogotá by Mr. Child: —

1. Glaucis hirsuta.
2. Phaëthornis emiliæ.
3. Phaëthornis anthophilus.
4. Campylopterus lazulus.
5. Lafresnaya flavicaudata.
6. Hyparoptila buffoni.
7. Florisuga mellivora.
8. Petasphora anais.
9. Petasphora cyanotis.
10. Panoplites flavescens.
11. Heliodoxa leadbeateri.
12. Pterophanes temmincki.
13. Docimastes ensiferus.
14. Helianthea typica.
15. Bourcieria torquata.
16. Floricola longirostris.
17. Heliotrypha exortis.
18. Thalurania columbica.
19. Acestrura mulsanti.
20. Acestrura heliodori.
21. Lesbia gouldi.
22. Lesbia amaryllis.
23. Cyanthus forficatus.
24. Rhamphomicron heteropogon.
25. Rhamphomicron microrhynchum.
26. Metallura tyrianthina.
27. Chrysuronia æuone.
28. Adelomyia melanogenys.
29. Aglæactis cupripennis.
30. Erioenemis alinæ.
31. Erioenemis cupriventris.
32. Erioenemis vestita.
33. Uranomitra franciæ.
34. Amazilia fuscicaudata.
35. Amazilia cyanifrons.
36. Hylocharis sapphirina.
37. Chlorostilbon angustipennis.
38. Panychlora poortmani.

It will always be a source of regret to me that before starting upon our trip I had not been able to obtain any information concerning the island of Curaçao other than that contained in the Encyclopædia and in the folder of the Red "D" Line. It is true that I did not make any great effort to this end, as at the time I expected that we would simply stop on the island between ships and have no opportunity to collect any birds. When we finally arrived and did have a chance to get some birds, we did not work them up as thoroughly as we should have done, because I could but think that an island so small, and under such perfect civilization and government for several hundred years, must be thoroughly explored and known. However, at the same time of our visit, Mr. Ernst Hartert of England was engaged upon the ornithology of the three islands, Curaçao, Aruba, and Bonaire, and he has subsequently published ("Ibis" for July, 1893) an article giving the results of his work. He announces several interesting discoveries, and to his article I should refer all who may desire a complete work upon the subject.

My observations make no definite additions to those of Mr. Hartert, with this exception, that is, that I took a specimen of the spotted sand snipe (*Actitis macularia*) which he mentions as having observed, but not taken.

My field notes are as follows: —

BIRDS OBSERVED ON THE ISLAND OF CURAÇAO.

1. *Sterna sp.* Large, dusky above, crown black.
2. *Sterna sp.* Medium size, apparently pure white.
3. *Sterna sp.* Very small.
4. PELECANUS FUSCUS (LINN). Brown Pelican.
   I saw several flocks flying over Santa Ana Harbor. Aud. *B. of N. A.* vol. 7, ppl. 423, 424.
5. FREGATA AQUILA (LINN.). Frigate Pelican.
   I saw one individual flying over the harbor. The native name is "tijereta," scissors, or scissor-tail. Aud. *B. of N. A.* vol. 7, pl. 421.
6. *Ardea virescens (Linn.)* ? Green Heron.
   I saw flying across the harbor several small herons which I took to be of this species. I saw others again in the mangrove swamp to the northwest of the town. Aud. *B. of N. A.* vol. 6, pl. 367.

7. ACTITIS MACULARIA (LINN.).   Spotted Sand Snipe.

I shot a specimen in the immature unspotted plumage on the edge of a salt-pan north of the town.   Along the southeast edge of the Lagoon I saw in July several species of sand snipe, but I did not have my gun with me and obtained no specimens. Aud. *B. of N. A.* vol. 5, pl. 342.

8. EUPSYCHORTYX CRISTATUS (LINN.).   Crested Partridge.

I obtained but one specimen of this partridge, though I saw a good many.   Three was the greatest number that I saw together. In June I saw half-grown young ones in captivity.   Gould, *Mon. of Odontophorinæ.*

9. COLUMBA GYMNOPTHALMA (TEMM.).   White-winged Pigeon.

I saw a young one in captivity in June, and in July I saw a flock of perhaps a dozen individuals, from which I obtained one specimen.

10. *Zenaida sp.*

I saw many doves of a medium size, but obtained no specimen.   They were probably *Z. vinaceo-rufa* (Ridgw.).

11. COLUMBIGALLINA PASSERINA (LINN.).   Ground Dove.

I found this little dove very abundant.   It was probably the commonest bird on the island. Aud. *B. of N. A.* vol. 5, pl. 283.

12. TINNUNCULUS SPARVERIUS BREVIPENNIS (V. BERL.). Curaçao Sparrow-hawk.

I saw a good many of these hawks.   There is the same difference in color between the sexes as in our species.   I was told that they fed on lizards.

13. A large hawk that I saw several times at a distance may have been *Buteo albicaudatus colonus* (V. Berl.).

14. CONURUS PERTINAX (LINN.).   Yellow-headed Parrakeet.

I saw many of these in captivity, some of them barely fledged, and was told that they were caught on the island, but I saw no others.

15. CHRYSOLAMPIS MOSCHITUS (LINN.).   Ruby and Topaz Humming-bird.

I saw none of these in June, but in July the tamarind-trees were in bloom, and there were swarms around every tree. Nearly, all, however, were in poor plumage, as they were just moulting. Gould's *Monograph*.

16. CHLOROSTILBON ATALA (LESS.). Atala Humming-bird.

I saw many of these in June and more in July, when they also were feeding on the tamarind blossoms. Gould's *Monograph*.

17. ICTERUS ICTERUS (LINN.). Troupial.

All of the troupials that I saw were caged birds, but I was told that the bird is found and breeds on the island. The natives apply the name "troupial" or "turupial" to both this and the following species, so that I cannot tell how much weight to give to my information. Aud. *B. of N. A.* vol. 7, pl. 499.

18. ICTERUS XANTHORNUS CURASOËNSIS (RIDGW.). Curaçao Oriole.

I saw a small flock of five or six in June, and in July I saw three more. Pl. in this work.

19. ZONOTRICHIA PILEATA (BODD.). Pileated Sparrow.

This handsome sparrow I found quite common in a little valley near the monastery. Descourtilz, *Ornithologie Brésilienne*.

20. EUETHEIA BICOLOR (LINN.). Grassquit.

We saw quantities of these wherever we went on the island. On July 26 we found a nest with three eggs.

21. CŒREBA UROPYGIALIS (V. BERL.). Curaçao Honey-creeper.

I saw but few in June, but in July they were abundant, and were seen in the tamarind-trees with the humming-birds. They have a feeble lisping song more like that of an insect than that of a bird.

22. DENDROICA RUFO-PILEATA (RIDGW.). Curaçao Warbler.

These were abundant. Their song is much like that of our yellow warbler.

23. MIMUS GILVUS ROSTRATUS (RIDGW.). Curaçao Mocking-bird.

These birds were abundant. I even saw some singing from the housetops in the town. Both of my specimens were young, and had speckled breasts. The native name is "ruiseñor," which is the Spanish for nightingale.

The barn-owl mentioned by Captain Smith was doubtless *Strix flammea bargei* (Hartert).

For the benefit of those who may not have access to Mr. Hartert's article, I give here his list of the birds of the island of Curaçao : —

1. Larus atricilla (Linn.).
2. Sterna hirundo (Linn.).
3. Sterna maxima (Bodd.).
4. Phalacrocorax brasilianus (Gm.).
5. Fregata aquila (Linn.).
6. Pelecanus fuscus (Linn.).
7. Hæmatopus palliatus (Temm.).
8. Himantopus mexicanus (Müll.).
9. Totanus macularius (Linn.).   Actitis macularia.
10. Butorides (Ardea) virescens (Linn.).
11. Ardea candidissima? (Gm.).
12. Ardea herodias (Linn.).
13. Eupsychortyx cristatus (Linn.).
14. Columbigallina passerina perpallida (Hartert). C. passerina.
15. Leptoptila verreauxi (Bp.).
16. Zenaida vinaceo-rufa (Ridgw.).
17. Columba gymnopthalma (Temm.).
18. Strix flammea bargei (Hartert).
19. Polyborus cheriway (Jacq.).
21. Tinnunculus sparverius brevipennis (V. Berl.).
21. Buteo albicaudatus colonus (V. Berl.).
22. Conurus pertinax (Linn.).
23. Crotophaga sulcirostris (Sw.).
24. Stenopsis cayennensis (Gm.).
25. Chlorostilbon caribæus (Lawr.).   C. atala.
26. Chrysolampis mosquitus (Linn.).

27. Tyrannus dominicensis (Gm.).
28. Sublegatus glaber (Scl. & Salv.).
29. Myiarchus brevipennis (Hartert).
30. Elainea martinica riisii (Scl.).
31. Hirundo erythrogastra (Bodd.).
32. Icterus icterus (Linn.).
33. Icterus xanthornus curasoënsis (Ridgw.).
34. Euetheia sharpei (Hartert).   E. bicolor.
35. Zonotrichia pileata (Bodd.).
36. Ammodramus savannarum (Gm.).
37. Certhiola uropygialis (V. Berl.).
38. Dendroica rufo-pileata (Ridgw.).
39. Mimus gilvus rostratus (Ridgw.).

CHAPTER IX.

I THINK that I may say without egotism that I can sometimes make a fair bird-skin, and the fact that others could do the same was to me, at one time, nothing remarkable, but now I regard with great respect the man who can go to the tropics and return with a collection of good skins.

Until we have experienced them ourselves, we do not realize the difficulties that beset the collector in the tropics. Suppose that you have come in at nightfall with ten or fifteen birds that you wish to save. You have your supper, and then you begin to realize that you are tired and sleepy, but still you start to work. You have a wretched spluttering tallow-dip for light, and mosquitoes come in clouds to harass you. However, you keep bravely on and finish one good skin, then look at your watch. You have, if the bird is a medium-sized one, been at work just twenty minutes; at this rate you will be four or five hours longer. The prospect is too much for you ; you make two or three more skins, then hang up the rest of the birds in the coolest place that you can find, and say that you will begin upon them at daybreak the next morning. When you wake, you at once notice a peculiar smell; you examine your birds; they are putrid, and must be thrown away at once. You still have the skins left, and later you take a look at them. You find them covered with thousands of little red ants, the skin of their feet and their eyelids have already been eaten off, and many feathers have been cut away, leaving unsightly bald patches. You take each skin, blow and dust off the ants, clean them thoroughly, and replace them

upon the drying board, which you suspend by strings. In less than
an hour the ants have found them again. You clean them a second
time, and now anoint the strings with carbolic acid, tar, kerosene
oil, and camphor, any of which you think would turn back an
insect with the slightest self-respect, but your trouble is for naught.
Later you find that water is the only thing that will keep them back,
so you borrow cans and plates, fill them with water, and arrange
pedestals upon which you think that your skins are safe. After a
while you hear a buzzing, you look at your skins, and see some
large green flies upon them. You drive them away, but the next
day you find gimlet-holes through the heads and beaks of your
birds. They were made by maggots hatched from eggs of the flies.
When at last your skins are dry, you go to pack them, and as you
lift them up, black beetles scurry out from under them. You find
that they have burrowed between the skin and bones of the tarsi
and wings of your skins, until they are mere shells ready to fall in
pieces at a touch. You collect up to the last day of your stay, and
have some green skins which you pack with the greatest care.
After two days on mule-back you arrive at your next station, and at
once proceed to air your skins. The green ones are dry enough
now; but what horrible monstrosities! — their necks twisted and bent,
their feathers lying the wrong way, their bodies distorted. Some of
your skins that you thought were thoroughly dried were evidently
not so, as they are now covered with a fungous mould, their black
beaks a pale silvery color crumbling at the touch. You resolve to
do better at this place, and you put your drying skins out in the
sun. In a little while you hear a noise, and look out in time to see
a black vulture flying up to the roof with your best skin, one that
you have taken especial pains to preserve. You stand helplessly
looking on until it is torn in pieces and left in disgust. After that,
you keep your skins indoors, but that night the mice take a fancy
to examine them, and the next morning you find the floor strewn
with wings and tails. It is but poor consolation to think that the
arsenic may perhaps have poisoned the mice.

All this may be thought overdrawn, but everything that I have related above occurred to me, and my object in writing this chapter is to point out to others who may hereafter go on similar expeditions how they may avoid my troubles.

First of all, I should advise you to collect, when possible, in the early morning. Get up by daybreak, and take a cup of coffee and a mouthful of food before going out. You will find all birds stirring at this hour, whilst at noon you will see few if any. Come in before eleven o'clock, then rest a little, have your breakfast, skin your birds carefully, write up your notes, prepare your ammunition, etc., for the next day, and go to bed early.

Now, in regard to your skins: once that a skin has been thoroughly dried in good shape it will stand packing and transportation with a fair amount of safety, therefore your aim should be to shape and dry them properly. You may travel with a large and specially prepared outfit, but small steamer trunks with several trays are very well suited for packing and drying skins, being lighter than chests, waterproof to some extent, and furnished with good locks.

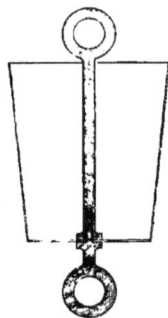

Skins, whilst drying, are undoubtedly safest from molestation when they are suspended, and are also more out of your way. A trunk-tray is easily hung up to a rope, a rafter, or a branch of a tree, your dried skins can be kept safely in it, and those that are drying can be placed on a board crosswise on the tray. The point comes up about the ants. I suggest the following: A couple of tin cups, through the bottoms of which pass a bar with a ring at each end. The

SECTION OF CUP.

tray is suspended from the lower ring, the cup in turn from the upper, and the cup being filled with water will effectually keep out the ants. I would also recommend that a piece of gauze or mosquito net be spread over the tray when it is suspended. This will not interfere with the drying, but will prevent the damage from the green flies and beetles. Skins, when dry and packed away in your trunk, must also be protected from ants, and the best way is to carry along three or four deep tin plates which will fit into one another for convenience in packing. Fill these with water, put a stone in each, and place your trunk on these stones.

In regard to your skinning outfit: that will depend upon your taste, but I recommend simplicity. I took a pair of small short-bladed, sharp-pointed scissors, a pair of tweezers, a pocket-knife, and a knitting-needle, and found this amply sufficient. A tool-handle, containing gimlets, screw-driver, small chisel, etc., was also very useful.

For materials: I was once in favor of plaster of paris in skinning, but I now prefer Indian-corn meal. Take it tied up in shot-bags. It is not heavy, will not spill out, and packs well in your trunk. When you are skinning, spread out a sheet of paper, and when you are through pour back the meal that is left. It can be used repeatedly.

Take cotton batting with you, the kind that is sold in our dry-goods stores done up in tissue paper. You can roll it up in an old towel, and by wrapping it tightly with a strong string can compress it until its size is many times reduced, and it packs away well. Cotton is found all over Colombia, and I thought that I was doing something unnecessary when I carried some with me; but in Gua-duas my supply gave out, and when I sent for some it was brought to me in little wads about the size of a walnut, and I found that I could no more stuff a bird with it than I could with a set of building blocks. However, it was owing to this that I became acquainted with what I consider a splendid material for stuffing large and medium-sized birds, — I refer to oakum as used for calking vessels. This is inelastic and retains the shape given to it, and a bird's body can be modeled exactly after the one of flesh just removed. Some birds have wide projecting shoulders, with a deep depression between the furculum and the neck, and it is just this shape that is difficult to stuff with cotton, but can be fitted like a glove with oakum. I obtained the oakum on the river steamer. The Spanish name for it is "estopa." The best preservative is dry arsenic, which should be carried in a tight can with a screw cap. The can should be conspicuously marked with both the English word "Poison" and the Spanish word "Véneno."

And now, to change the subject abruptly, I would say a few words about photography. The ability to sketch rapidly and accurately is much to be envied, yet for one person with Mr. Catherwood's talent, there are ten thousand with-

RED-TAILED HAWK (LIVING BIRD).

out; and then, too, how rapidly a camera does its work! I believe a camera to be nowadays an essential part of every traveler's outfit; yet it has its limitations. In regard to choice of cameras, it is like choice of shot guns; every one thinks his own the best. I took with me a "Hawk-eye," taking 4 × 5 plates, and used glass plates entirely, which were developed upon my return. Many of the preceding illustrations are from my photographs. I purchased my camera several years before for the purpose of taking pictures of objects of natural history; and it is of this class of work that I wish to speak.

Considering birds first, although I am aware that some students have taken fine pictures of them, I have not met with success. Those pictures that I have taken of birds in a state of freedom have not turned out well, usually because of the smallness of the figure and of the impossibility of selecting a suitable background. I have had wounded birds from time to time; but it is very difficult to

get them in a good position; and the best pictures are but poor.
When seriously wounded, their listless and dejected look is not
what is wanted in a picture, and at the best the surroundings of
fences, cages, chains, or cords destroy the worth of the likeness.
A dead bird is a hopeless task, and photographs of stuffed and
mounted birds, with their dull, protuberant, and lifeless eyes are an
abomination. I will venture to say that not one photograph in a
hundred of mounted birds has the faintest life-like look about it.

Leaving birds and turning to fish, we find a class that, as a rule,
make elegant subjects for the photographer. With them I have
been quite successful. The best to work upon are scale fish of
moderate size; but I have made good pictures of sharks of five
feet in length and of small fish of barely three inches. I would

GREEN HERON (MOUNTED SKIN).

recommend this work to lovers of photography, and will give a
brief explanation of my process. You will need a large sheet of
white blotting-paper, some small wire nails and pins, and a pair of
wire-cutting pliers. Select a moderate-sized fish, with uninjured
tail and fins, fasten your blotting-paper to a board, wipe the fish
dry, and lay it in the centre of the sheet of paper. Cut off the
heads of two of the wire nails, and drive them through the fish and
into the board, one near the head, and the other near the tail.

Drive the nails until they are below the level of the skin, when the

WHITE PERCH.

scales will cover the holes, and they will not be seen. These will support the weight of the fish. Now, with your pliers cut off the heads of a number of pins, and use the points to keep spread the fish's tail and fins. Put the board on edge, and move your camera up until the fish nearly covers the plate. Use a very slow plate, and give plenty of exposure. When you have developed your plate, and come to print it, vignette closely to the fish, by which means you can get an almost dead white background. In case you do not use a white background, you may still print on aristotype paper, and then, with a sharp eraser, scratch off all the print except the fish, thus getting a pure white background. This sea bass, a male, with the dorsal

SEA BASS.

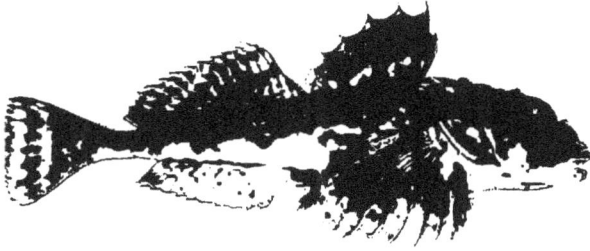

SCULPIN.

hump, characteristic of the breeding season, printed well; but its tail was badly cut up in the net.

I believe this method would be useful to travelers who have not with them the means of preserving specimens of strange fish that they may see. It might enable them to identify these fish upon

SCULPIN.
(By Permission.)

their return. The red snapper figured on page 148 is from a photograph taken as described above. That this method is also applicable to fish without scales, the figure of a sculpin will show, and that the comparative value of the illustration can be judged, I give also a wood-cut of a sculpin, from a popular work on natural history.

Crustacea may be treated in the same manner as fish, and the results are equally as satisfactory. The various crabs found along our coast are good subjects upon which to work.

LOBSTER.

Some insects may be successfully photographed; but here the question of color is so difficult to deal with that one must be an expert before he can count upon the result. The black and yellow butterflies especially are disappointing to handle. What I have said of birds applies with even more force to animals, and to a less degree to reptiles. The colors of snakes and of terrapin are lost, and of the other reptiles, it is indeed rarely that one can be gotten to take a good attitude, and hold it long enough for a good picture.

APPENDIX.

— —

LIST OF WORKS ON COLOMBIA.

A LIST of the works treating of any particular subject is always of great help to the student who may care to investigate that subject, and therefore I have compiled the following list of works on Colombia, exclusive of the literature of the Isthmus of Panama. The nature of many of these works is indicated by their titles; in others this is not the case, and as I have not had access to many of them, nor time to read others, I have been compelled to adopt a chronological arrangement. I have, however, collected the purely zoölogical writings and brought them together after the general list.

It is to be borne in mind that in early days Colombia included at various times more or less of Peru, Ecuador, and Venezuela, and was known by other names as Tierra Firma, New Granada, etc. Examining the titles below, it will be seen that the earlier writings were those of the Jesuits; then came in historical and biographical works. In the early twenties, during the struggles of Colombia in securing her independence from Spain, many officers of foreign armies were attracted as adventurers, and later a number of them wrote of their experiences. Then followed a period in which little appeared. Within the last fifteen years there has been a large increase in the literature on Colombia, in which the Germans have had a prominent part. In this compilation I have profited by Pereira's list (see No. 155). In Bonnycastle's work, 1819, there is a list of 146 works on Spanish America, many of which may have references to Colombia, but as I have not seen them I cannot include them.

1. De insulis nuper inventis. Occeanea decas. Petrus ab Angleria Martir. Legatio Babilonica. Poemata. Seville, 1511.

2. La cronica del Peru. Pedro Cieza de Leon. Amberes. 1554.

3. Elegias de varones ilustres de Indias. Juan de Castellanos. Madrid. 1589.

4. Historia de las Indias y Cronica de la Nueva España. Francisco Lopez Gomara. Madrid. 1600.

5. Histoire naturelle et morale des Indes; tant orientales qu'occidentales, par

178   *APPENDIX.*

le P. Joseph de Acosta; traduite en françois par R. Regnault Cauxois.   Paris, 1600.

6. Grammatica en la lengua del nuevo reyno llamada Mosca.   Bern. de Lugo. Madrid, 1619.

7. Noticias historiales de las conquistas de Tierra Firme en las Indias Occidentales.   Fray Pedro Simon.   Cuenca, 1627.

8. Varones ilustres del nuevo mundo, descubridores, conquistadores y pacificadores del opulento, dilatado y numeroso imperio de las Indias Occidentales.   Fernando de Pizarro.   Madrid, 1639.

9. L'Histoire du nouveau monde, ou Description des Indes Occidentales.   Le sieur Jean de Laet.   Leyden, 1640.

10. Arte y vocabulario de la lengua de los Indios de la provincia de Cumana o Nueva Andaleucia.   Fr. de Tauste.   Madrid, 1680.

11. Historia general de la conquista del nuevo reyno de Granada.   Lucas Fernandez de Piedrahita.   Madrid, 1688.

12. Historia general de los hechos de los Castellanos en las islas i tierra firme del mar Occeano.   Antonio de Herrera.   Madrid, 1729.

13. El Orinoco ilustrado, historia natural, civil y geografica de este gran rio y de sus caudalosas vertientes.   Le P. Joseph Gumilla de la Compagnie de Jesus. Madrid, 1741.

14. Historia de la Provincia de Santa Fe, de la compañia de Jesus y vidas de sus varones ilustres.   Le P. Joseph Cassani.   Madrid, 1741.

15. Journal du voyage fait par ordre du roi à l'Equateur.   M. de la Condamine. Paris, 1751.

16. Historia coro-graphica, natural y evangelica de la Nueva Andalucia.   Provincias de Cumaná, Guayana y Vertientes del Rio Orinoco.   Fr. Antonio Caulin. Madrid, 1779.   One quarto vol., pp. 482+13, engraved title, and three plates of murders of priests by Indians.

17. A philosophical and political history of the settlements and trade of the Europeans in the East and West Indies, by the Abbé Raynal, translated by J. O. Justamond.   London, 1783.   Eight vols., many maps.   (Santa Marta, Cartagena, and Santa Fe de Bogotá in 4th vol., pp. 58 to about 105 incl.)

18. La perla de la America, provincia de Santa Marta.   Antonio Julian.   Madrid, 1786.

19. Diccionario geographico-historico de las Indias Occidentales o America.   Col. Antonio de Alcedo.   Madrid, 1786–88.

20. Historia del Nuevo Mundo.   Juan Bautista Muñoz.   Madrid, 1793.

21. Voyage à la partie orientale de la Terre-Ferme.   Depons.   Paris, 1806.

22. A voyage to South America: describing at large the Spanish cities, towns, provinces, etc., on that extensive continent: undertaken by command of the king of Spain.   By Don George Juan, and Don Antonio de Ulloa.   Translated from the Spanish by John Adams.   London, 1807.   Two vols., pp. 28+479 and 4+419 and index; many maps and plates.   (Cartagena in first vol., pp. 19–84.)

23. Voyage dans l'intérieur de l'Amérique dans les années 1799 à 1804. Par A. de Humboldt et A. Bonpland. Paris, 1807–39. Six parts, 1494 plates (349 colored), 5 maps.

There are many other editions of Humboldt's writings.

24. Spanish America; or a descriptive, historical, and geographical account of the Dominions of Spain in the Western Hemisphere, continental and insular. R. H. Bonnycastle. Philadelphia, 1819. One vol., pp. 482. (New Granada, pp. 159 to 240 incl. Map.)

25. Barthélemi Casas. Evêque de Chiapa: Œuvres précédées de sa vie. Paris, 1822.

26. Colombia: being a geographical, statistical, agricultural, commercial, and political account of that country, adapted for the general reader, the merchant, and the colonist. London, 1822. Two vols., pp. 124+707 and 782, two portraits, one large folding map. This is known as Walker's Colombia. There is also a Spanish edition.

27. De Republiek Columbia, of Tafereel Van Derzelver Tegenwoor. digen toestand en Betrekkingen; in Brieven, van daar aan zijne vrienden geschreven, door Carl Richard, Hanoversch officier. Benevens eene levensschets van Simon Bolivar, President van Columbia. Amsterdam, 1822. One vol., pp. 285.

28. The geography, history, and statistics of America, and the West Indies; exhibiting a correct account of the discovery, settlement, and progress of the various kingdoms, states, and provinces of the Western Hemisphere, to the year 1822. By H. C. Carey and I. Lea. Philadelphia. With additions relative to the new states of South America. London, 1823. One vol., pp. 477, three folding maps. (Colombia, pp. 412 to 423 incl.)

29. Letters written from Colombia, during a journey from Caracas to Bogotá, and thence to Santa Martha, in 1823. London, 1824. One vol., pp. 16+208, one large folding map.

30. Colombia: its present state, in respect of climate, soil, productions, population, government, commerce, revenue, manufactures, arts, literature, manners, education, and inducements to emigration: with an original map: and itineraries, partly from Spanish surveys, partly from actual observation. By Col. Francis Hall. London, 1824. One vol., pp. 6+154, one map.

31. Voyage dans la République de Colombia, en 1823. G. Mollien. Paris, 1824. Two vols., pp. 4+308 and 316, seven colored plates, and one large folding map.

32. Travels in the Republic of Colombia, in the years 1822 and 1823. By G. Mollien. (Translation of the above.) London, 1824. One vol., pp. 460, one plate, one folding map.

33. Journal of a residence and travels in Colombia, during the years 1823 and 1824. By Capt. Charles Stuart Cochrane, R. N. London, 1825. Two vols., pp. 16+524 and 8+517, two colored plates, one large folding map.

34. Coleccion de los viajes y descubrimientos que hicieron por mar los Españoles desde fines del siglo XV. Martin Fernandez de Navarrete. Madrid, 1825–29.

35. A visit to Colombia in the years 1822 and 1823, by Laguayra and Caracas, over the cordillera to Bogotá, and thence by the Magdalena to Cartagena. By Col. Wm. Duane. Philadelphia, 1826. One vol., pp. 632, two plates.

36. Histoire de la Colombie, par M. Lallement. Paris, 1826, second edition. One vol., pp. 320, one folding map and plates.

37. Notes on Colombia, taken in the years 1822–3, with an itinerary of the route from Caracas to Bogotá; and an appendix. Capt. Richard Bache, U. S. A. Philadelphia, 1827. One vol., pp. 303, two folding maps, one plate.

38. Travels through the interior provinces of Columbia. By Col. J. P. Hamilton. London, 1827. Two vols., pp. 332 and 256, seven plates, one map. Contains many references to birds and shooting.

39. Recollections of a service of three years during the war-of-extermination in the republics of Venezuela and Colombia. By an officer of the Colombian navy. London, 1828. Two vols., pp. 15 + 251 and 8 + 277.

40. Die Geschichte von Columbia, durch Dr. Ernst Munch. Dresden, 1828. Two vols., pp. 113 and 111.

41. Colombia in 1826. By an Anglo-Colombian. In the Pamphleteer, vol. 29, London, 1828. pp. 485–505.

42. History of the life and voyages of Columbus. Washington Irving. 1828. Three vols.

43. The Modern Traveller: a popular description, geographical, historical, and topographical, of the various countries of the globe. Colombia. Vol. viii. Boston and Philadelphia, 1830. Pp. 336, three plates, one folding map.

44. Resa i Colombia, åren 1825 och 1826, af Carl August Gosselman Lieutenant vid Kongl. Maj:ts flotta. Stockholm, 1830. Two vols., pp. 274 and 300, two plates, one folding map.

45. The companions of Columbus. Washington Irving. 1831.

46. Sur la cause qui produit la goitre dans les cordillères de la Nouvelle-Granada. Boussingault, in Annales de Chimie, vol. 48. 1831. p. 41 et seq.

47. Sur les salines iodifères des Andes. Boussingault, in same, vol. 54, 1833, p. 163 et seq.

48. Comunicaciones entre el Señor Carlos Biddle, Coronel de los E. Unidos del Norte I la Sociedad Amigos del Pais. Panamá. 1836. Pamphlet, pp. 22, and one folding profile.

49. Antiguedades neo-granadinas. Ezequiel Uricoechea. Leipzig, 1837.

50. Voyages, relations et mémoires originaux pour servir à l'histoire de la découverte de l'Amérique. Ternaux-Compans. Paris, 1837–1841.

51. History of the reign of Ferdinand and Isabella the Catholic. W. H. Prescott. Three vols., portraits, maps, etc. 1838.

52. Bogotá in 1836–7, being a narrative of an expedition to the capital of New-

Grenada, and a residence there of eleven months. By J. Steuart. New York, 1838. One vol., pp. 312.

53. L'Univers. Histoire et Description de tous les peuples. Colombie et Guyanes, par M. C. Fanin. Paris, 1839. Pp. 32, one folding map, and seven plates.

54. Geografia historica, estadistica y local de la provincia de Cartagena. General Juan Jose Nieto. Cartagena. 1839.

55. Beiträge zur geologie von Antióquia; and Über die Salzquellen des nördlichen Theiles der Provinz Antióquia und die Gebirgs-Formationen der Umgebung von Medillin im Freistaate von Neu-Granada. C. Degenhardt, in Karsten's Archiv. für mineralogie, xii., 1839, p. 1 et seq.

56. Resumen de la geografia de Venezuela. Augustin Codazzi. Paris, 1841.

57. Resumen de la historia antigua de Venezuela. Baralt y Diaz. Paris, 1841.

58. Essai sur l'ancien Cundinamarca. Ternaux-Compans. Paris, 1842.

59. History of the conquest of Mexico. W. H. Prescott. 1843. Three vols., three portraits, two maps.

60. Vidas de los Españoles celebres. Manuel Jose de Quintana. Paris, 1845.

61. History of the conquest of Peru. W. H. Prescott. 1847. Two volumes, portraits and map.

62. Compendio histórico del descubrimiento y colonizacion de la Nueva Granada en el siglo décimo sexto. Col. Joaquin Acosta. Paris, 1848. One vol., pp. 460, four plates.

63. Semanario de la Nueva Granada: miscelanea de ciencias, literatura, artes e industria. Francisco Jose de Caldas. Edition of Col. Acosta. Paris, 1849.

64. Viajos cientificos á los Andes ecuatoriales. Boussingault. Paris, 1849. Pp. 67.

65. Coleccion de memorias sobres fisica, quimica e historia natural de la Nueva Granada y Ecuador, escritas por M. Boussingault, actual presidente de la Academia de Ciencias de Paris; traducidas con anuencia del autor y precedidas de algunas nociones de geologia. Col. Joaquin Acosta. Paris, 1849.

66. Observations diverses sur les environs de Santa Fé de Bogotá. P. A. Cornette, in Bulletin de la Société géologique de France. Second series, vol. 7, 1849-50, p. 320.

67. Memorias para la historia de la Nueva Granada, desde su descubrimiento hasta el 20 de Julio de 1810. Col. José Antonio de Plaza. Bogotá, 1850.

68. Acosta. Sur les montagnes trachytiques de Ruis, dans la Nouvelle Grenade. In Bull. de la Soc. Géol. de France, Paris, 1850-51, pp. 489 to 496 incl., one plate of maps, sections, etc.

69. Same. Sur la Sierra Névada de Sainte-Marthe. Formée par le terrain primitif. In same for 1851-52, pp. 396 to 399 incl., one folding plate of sections.

70. Extrait de differentes lettres sur la géologie de la Nouvelle Grenade. P. A. Cornette, in same for 1851-52, p. 509.

71. Geognostische Bemerkungen über die nord küste Neu-Granada's, insbesondere über die sogenannten vulkane von Turbaco und Zamba. H. Karsten, in Zeitschrift der Deutschen Geologischen Gesellschaft, 1852. p. 579.

72. Memoria sobre la geografia de la Nueva Granada. Mosquera. New York, 1852.

73. Memoir on the physical and political geography of New Granada. General T. C. de Mosquera. Translated from the Spanish by Theodore Dwight. New York, 1853. One vol., pp. 105, one large folding map. (Translation of the preceding.)

74. Peregrinacion de Alpha (M. Ancizar.) por las provincias del norte de la Nueva Granada, en 1850 I 51. Bogotá, 1853. One vol., pp. 524, portrait.

75. Resúmen histórico de los acontecimientos que han tenido lugar en la república, extractado de los diarios y noticias que ha podido obtener el general gefe del estado mayor general, T. C. de Mosquera. Bogotá, 1855. One vol., pp. 226 + 74.

76. Jeografia fisica I politica de las provincias de la Nueva Granada, por la Comision Corografica. Provincias del Socorro, Velez, Tunja I Tundama. Bogotá, 1856. One vol., pp. 363.

77. Ueber die geognostischen Verhältnisse der Westl. Colombia. Karsten. Vienna, 1856.

78. Physiognomie der Trop. Vegetation Süd Americas. Albert Berg. 1856. Folio views of Colombian scenery.

79. History of the reign of the emperor Charles the Fifth. Wm. Robertson, W. H. Prescott. 1857. Three vols., portrait.

80. New Granada: Twenty months in the Andes. Isaac F. Holton, M. A. New York, 1857. One vol., pp. 605, 33 woodcuts, two colored double-sheet maps.

81. Positions bestimmungen und Höhenmessungen in Süd Amerika. Von Liais und Friesach. Sitzungsberichte der K. K. Akademie der wissenschaften, mathemat., naturw. No. 19. pp. 285–328, No. 38, pp. 591–632, No. 93, pp. 7–14; years 1857, '59, '60.

82. Historia de la revolucion de la República de Colombia en la América Meridional, por José Manuel Restrepo. Besanzon, 1858. Four octavo vols. First edition was in 1827.

83. Geografia de la República del Ecuador. Manuel Villavicencio. New York, 1858.

84. Beiträge zur geologie des Westl. Columbien. Karsten. Amtl. Bericht der Wiener Naturforscherversammlung. 1858.

85. The West Indies and the Spanish Main. Anthony Trollope. London, 1859. One vol., pp. 395, map. (New Granada, pp. 242–255.)

86. Voyage aux Indes Occidentales. Anthony Trollope, 1858–59. Dessins inédits par M. A. de Berard. In Tour du Monde, vol. 2, pp. 49 to 64 incl., one map, 8 cuts. (From preceding.)

87. Mapoteca Colombiana. Coleccion de los títulos de todos los mapas, planos, vistas, etc., relativos á la América española, Brasil e islas adyacentes; arreglada

cronologicamente y precedida de una introduccion sobre la historia cartográfica de América. Ezequiel Uricoechea. London. 1860.

88. Antiquarian, ethnological, and other researches in New Granada, Equador, Peru, and Chile, with observations on the pre-incarial, incarial, and other monuments of Peruvian nations; with plates. William Bollaert. London. 1860.

89. Lieutenant Michler's report of his survey for an inter-oceanic ship canal near the Isthmus of Darien. Atrato River survey. Senate Document. Feb., 1861. 1st vol., pp. 457; annotated list of 144 species of birds by Cassin, pp. 220–254. 2d vol. contains 17 large folding maps and profiles.

90. Ensayo sobre las revoluciones politicales y la condicion de las repúblicas Colombianas. Samper. Paris, 1861.

91. Dr. Moriz Wagner, in Petermann's Mittheilungen for 1861.

92. Same, in same for 1862. Eine Reise in das Innere der Landenge von San Blas und der Cordillere von Chepo in der Provinz Panama, mit besonderer Berücksichtigung der hypsometrischen verhältnisse und der Kanal frage. pp. 128–141, colored map.

93. Jeografia fisica I politica de los Estados Unidos de Colombia. Felipe Perez. Bogotá. 1862. Two vols., pp. 13 + 494 and 4 + 650, 8 plates.

94. Jeografia fisica I politica del Distrito Federal, Capital de los Estados Unidos de Colombia. Felipe Perez. Bogotá. 1862. One vol.. pp. 54.

95. Anales de la revolucion de 1861. Felipe Perez. Bogotá. 1863.

96. New Granada; its internal resources. Powles. London, 1863.

97. Vida del Libertador Simon Bolivar. Felipe Larrazabal. New York, 1865–75.

98. Compendio de geografia de los Estados Unidos de Colombia. Mosquera. London, 1866.

99. Autobiografia del General Jose Antonio Paez. New York, 1867.

100. Historia de la literatura en Nueva Granada. Vergara y Vergara. Bogotá, 1867.

101. Historia eclesiastica y civil de la Nueva Granada. Jose Manuel Groot. Bogotá. 1868–71.

102. Relaciones de los Vireyes del Nuevo Reino de Granada. Garcia y Garcia. New York, 1869.

103. Deutsche Konsulatsberichte aus Bogotá im Preussischen Handelsarchiv, 1870–75.

104. Informe de los esploradores del Territorio de San Martin. Bogotá, 1871. One vol.. pp. 4 + 59.

105. Esploracion entre San José de Cucutá I el Rio Magdalena. Bogotá. 1871. One vol.. pp. 18.

106. Alturas tomadas en la Republica de Colombia, en los años de 1868 y 1869. por W. Reiss y A. Stübel. Quito. 1872. One vol., pp. 39, principally barometric heights.

107. Voyage à la Nouvelle Grenade, par M. le Docteur Saffray. 1869. In Tour du Monde, vols. 24, 25, and 26, 1872–73, total pp. 160 and 110 cuts, one double-page map.

108. Memorias del General Joaquin Posada Gutierrez. Bogotá, 1872–80.

109. Höhenmessungen in Süd-America. In Zeitsch. der Gesell. für Erdkunde zu Berlin, 1874. pp. 440, 441.

110. Historia económica y estadistica de la Hacienda nacional. Anibal Galindo. Bogotá, 1874.

111. Genealogias del nuevo reino de Granada. Juan Florez de Ocariz. Madrid, 1874.

112. Compendio de historia patria. José Maria Quijano Otero. Bogotá, 1874.

113. Biografias militares. José Maria Baraya. Bogotá, 1875.

114. Die Culturlander des alten Amerika. Prof. Bastian. Berlin, 1875–76. 2 starken banden. pp. 720–1005, 3 maps.

115. Memorias de un abanderado, 1810–1819. José Maria Espinosa. Bogotá, 1876.

116. Dr. Reiss and Dr. Stübel. Höhenmessungen in den Republiken Colombia und Ecuador. Zus. ammengestellt von Prof. Meinicke. XII. Jahresbericht des Vereins für Erdk. zu Dresden. 1876.

117. Barometrische Höhenbestimmungen in Columbien von Eduard Steinheil, in Petermann's Mitt. for 1876, No. 8. pp. 281–284, colored folding map.

118. Reisen in Columbien von Eduard Steinheil. in Petermann's Mitt., 1876, No. 10, pp. 393–395, 1877. No. 4. pp. 184–188. No. 6, pp. 222–227.

119. Reise durch den Staat-Magdalena in Colombia, 1874. Tetens, in Mitt. geogr. gesell., Hambourg. 1876–77, pp. 367–70.

120. My first trip up the Magdalena, and life in the heart of the S. Andes. J. A. Bennett, late U. S. Consul at Bogotá, in Journal of the Amer. Geog. Soc. of New York. 1877. pp. 126–141.

121. L'Amérique équinoxiale: Colombie, Equateur, Pérou, par M. Ed. André, 1875–76. In Tour du Monde for 1877, '78, '79, and '83, total pp. 384, 286 cuts, 16 maps.

122. L'Amérique du Sud; voyage dans la Nouvelle Grenade. E. André, in L'Exploration, 1877, No. 20.

123. Reisen in nordvestlichen Sud-America. E. André, in Globus, 1878.

124. Anales diplomaticos de Colombia. Pedro Ignacio Cadena. Bogotá, 1878.

125. Recuerdos históricos 1819–1826. Coronel Manuel Antonio Lopez. Bogotá, 1878.

126. Reisen in Süd-America, 1868–1877. Reiss and Stübel, in Petermann's Mitt. for 1878, pp. 30–33.

127. Reiseerlebnisse in Columbien. W. Petersen, in Sitzungsber, naturforscher gesell. in Dorpat. 1878. pp. 42–47.

128. Les Chibchas de la Colombie. E. Uricoechea. Congr. intern. de science géogr. Paris, 1878, pp. 310–315.

129. Annotations sur les quinquinas des Etats-Unis de Colombie. D. E. Coronado. Paris, 1878, pp. 55.

130. Die Kulturländer des alten Amerika. A. Bastian. Berlin, 1878.

131. Über entdeckungen in Süd-Amerika. Bastian, in Verhandl. Ges. f. Erdk. Berlin, 1878, pp. 144–147.

132. Die Zeichen-Felsen Columbiens. Bastian, in Zeits. Ges. f. Erdk. Berlin, 1878, pp. 1–23.

133. Travels in Columbia in 1875–76. Ed. André, in Bull. de la Soc. de Géog. de Paris, 1879.

134. Edelmetall produktion und werthverhältnisse zwischen gold und silber, seit der entdeckung Amerika's vis zur gegenwart. Dr. Adolph Loetbeer, in Petermann's Mitt. Ergänzungsheft, No. 57, 1879. (Neu-Granada, pp. 60–64.)

135. Compendio de historia de Hispano-América. Cesar C. Guzman. Paris, 1879.

136. Diccionario biográfico de los campeones de la libertad de Nueva Granada, Venezuela, Ecuador y Peru. Leonidas Scarpetta, Saturnino Vergara. Bogotá, 1879.

137. Galeria nacional de hombres ilustres o notables. Jose Maria Samper. Bogotá, 1879.

138. Diccionario Jeográfico de los Estados Unidos de Colombia, por Joaquin Esguerra Ortiz. Bogotá, 1879. One vol., pp. 284.

139. Notes on the topography of the Sierra Nevada of Santa Marta, U. S. of Colombia, by F. A. A. Simons, in P. R. G. S. of London for Nov., 1879, pp. 689–694, one folding map.

140. Memorias del General Daniel Florencio O'Leary. Caracas, 1879–81. 16 vols.

141. Bosquejo estadistico de la region oriental de Colombia. Joaquin Diaz Escobar. 1880.

142. Historia de Colombia, contada a los Niños. José Joaquin Borda. Zipaquirá, 1880.

143. Explorations aux Isthmes de Panama et de Darien en 1876–77–78, par M. A. Réclus. In Tour du Monde for 1880, pp. 321–400. 68 cuts, 2 maps.

144. Reisen in Antióquia. Friederich von Schenck, in Petermann's Mitt. for 1880, pp. 41–47, large folding map.

145. Colombia e Peru, l'imperio degli Inca. G. B. Lemoyne. Turin, 1880.

146. Quelques mots sur la géologie de l'Etat d'Antióquia. Petitbois, in Annal. Soc. Géolog. de Belgique, 1880, pp. 159–163.

147. Voyage à la Sierra-Nevada de Sainte-Marthe. Elisée Réclus. Paris, 1881. One vol., pp. 6+337, 21 cuts, one folding map. The first edition was in 1861.

148. Los communeros : historia de la insurreccion de 1781. Manuel Briceno. Bogotá, 1881.

149. La mission de la Goajira, Nouvelle Grenade. Jannsen, in Les missions catholiques. 1881. No. 627.

150. Voyage sur le Rio Magdalena à travers les Andes et sur l'Orénoque. J. Crévaux, in Bull. Soc. Géogr. de Paris, July, 1881. pp. 7–25, with map.

151. On the Sierra Nevada of Santa Marta and its watershed. (State of Magdalena, U. S. of Colombia.) F. A. A. Simons, in P. R. G. S. of London for Dec., 1881. pp. 702–723, one folding map.

152. Über Francisco de Cáldas, den Neu-Granadinischen naturforscher und geographer. A. Schumacher, in Verhandl. d. Gesell. f. Erdk. zu Berlin, 1881.

153. Voyage d'exploration à travers la Nouvelle Grenade et le Vénézuela (Rios Magdaléna, de Lesseps ou Guaviare, Orinoco). J. Crévaux et E. Lejanne. 1881. In Tour du Monde for 1882, pp. 225–320, 68 cuts, 2 maps.

154. El Dorado. Illustrated periodical, Liborio Zerda. Bogotá, 1882.

155. Les Etats-Unis de Colombie : precis d'histoire et de géographie physique, politique et commerciale, etc., etc., etc. R. S. Pereira. Paris, 1883. One vol., pp. 8 + 311, 10 double-page maps, one folding.

156. Geografía general de los Estados Unidos de Colombia. Bogotá, 1883. Pp. 456.

157. Voyage à la Nouvelle Grenade. Lejanne, in Bull. Soc. Géogr. Brest, 1883.

158. Ferrocarril de Antióquia : Informe de una comision. Medillin, 1883. Pp. 33.

159. Reiseskizzen aus Columbia. Hettner, in Kölnische Zeitung, 1883.

160. Notes on the central provinces of Colombia. Robert Blake White, in P. R. G. S. of London for May, 1883. pp. 249–267, one folding map.

161. Reisen in Antióquia im jahre 1880 von Fr. von Schenck.

162. Reisen in Antióquia und im Cauca im jahre 1880 und 1881. Fr. von Schenck.

163. F. v. Schenck's Höhenmessungen in Kolumbien, von Professor K. Zöppritz.

164. Hohen. in Antióquia nach White. All in Petermann's Mitt., 1883. Folding maps.

165. Voyages et découvertes de J. Crévaux. G. Franck. Paris, 1884. Pp. 88, maps.

166. Mapa para servir de estudio de la frontera entre Venezuela y Colombia. Rivadeneyra. Madrid, 1884. 5 maps.

167. Die Republik der Vereinigten Staaten von Kolumbien. W. Roth, in Das Ausland, 1884.

168. Spanish and Portuguese South America during the Colonial Period. R. G. Watson. London, 1884. Two vols., pp. 620.

169. Südamerikanische studien, drie Lebens und Kulturbilder. Mútis. Cáldas. Codazzi. H. A. Schumacher. Berlin, 1884. Pp. 559.

170. Estudio sobre las minas de oro y plata de Colombia. Vicente Restrepo, in An. de la instruccion pública, 1884.

171. Geografía general y compendio historico del Estado de Antióquia en Colombia. Manuel Uribe Angel. Paris, 1885. One vol., pp. 15 + 783. frontispiece, 33 cuts of antiquities, two folding maps.

172. Les anciennes populations de la Colombie. Nadaillac. Paris, 1885. Pp. 13.

173. Notes ethnographiques sur les Etats Unis de la Colombie. W. Boye, in Rev. Soc. Géogr. Tours, 1885.

174. Die Sierra Nevada von Santa Marta. Hettner, in Petermann's Mitt. for 1885. pp. 92–97.

175. The Sierra Nevada of Santa Marta. Sievers, in Proc. Geog. Soc. of Berlin, 1885.

176. Die barometrischen höhenmessungen des Herrn Dr. Sievers in Columbia und Venezuela. M. Frohberg, in Mitt. Geogr. Ges. Hamburg. 1885–86.

177. An exploration of the Goajira Peninsula, U. S. of Colombia. F. A. A. Simons, in P. R. G. S. of London for Dec., 1885. pp. 781–796. one folding map.

178. Reise nach Bogotá: Haupstadt der südamerikanischen Republik Colombia. E. Rothlisberger, in Jahresber. d. Geogr. Ges. Bern. 1885–87.

179. Stati Uniti di Colombia. Segre. Rome. 1886.

180. Costa Rica y Colombia de 1573 á 1881 : su jurisdiccion y sus limites territoriales. De Peralta. Madrid y Paris, 1886.

181. Géologie de l'ancienne Colombie Bolivarienne ; Vénézuela, Nouvelle-Grenade et Ecuador. Karsten. Berlin, 1886. Maps and ppl.

182. A study of the gold and silver mines of Columbia. V. Restrepo. Translated by C. W. Fisher. New York. 1886.

183. Die wirtschaftlichen verhältnisse der Vereinigten Staaten von Kolumbien. Hettner, in Dtsch. Kolonialzeitung, 1886.

184. Die Bogotaner. Hettner, in Globus, 1886.

185. Cartagena y sus cercanías : guia descriptiva. J. P. Urueta. Cartagena, 1886.

186. Die Arhuaco Indianer in der Sierra Nevada de Santa Marta. Sievers, in Zeits. Ges. f. Erdk. Berlin, 1886.

187. Reise in der Sierra Nevada de Santa Marta. Sievers, in same. 1886.

188. Über ein Skelett und schädel von Goajiros. Virchow, in Ges. f. Anthropol., etc. Berlin, 1886.

189. Historia del Nuevo Reino de Granada. Juan de Castellanos. Madrid, 1886–87.

190. Historia de Colombia. Carlos Benedetti. Lima, 1887. One vol. pp. 961.

191. Nouvelle-Grenade : Apercu général sur la Colombie et récits de voyages en Amérique. C. P. Etienne. Geneva. 1887. Pp. 144.

192. Descripcion histórica, geográfica y política de la República de Colombia. Bogotá, 1887. Pp. 23.

193. Compendio de geografia de la República de Colombia. A. M. D. Lemos. Medellin, 1887.

194. The U. S. of Colombia and the Isthmus of Panamá. J. Xantus, in Bull. Soc. Hongr. de Geogr.. 1887.

195. The Sierra Nevada de Santa Marta. J. T. Bealby, in Scottish Geogr. Mag., 1887.

196. Reise in der Sierra Nevada de Santa Marta. Sievers. Leipzig. 1887. Pp. 10 + 290.

197. The Goajira Peninsula: trade. etc. E. H. Plumacher, in Reports U. S. Consuls. 1887.

198. Ethnogr. stellung der Guajiro Indianer. A. Ernst, in Zeits. f. Ethnologie, 1887.

199. Briefe aus Kolumbien. F. C. Lehman, in Export, 1887.

200. The agricultural condition of Columbia. Wheeler, in Diplom. and Consular Reports. London, 1887.

201. Circulaire du ministre des affaires étrangères sur les mines d'or et d'argent de la République de Colombie, 1887.

202. The capitals of Spanish America. Wm. E. Curtis. New York, 1888. (Bogotá, pp. 225–257, 18 cuts.)

203. Reisen in den Columbianischen Anden. Dr. Alfred Hettner. Leipzig, 1888. One vol., pp. 10 + 389, one folding map.

204. Kartographische ergebnisse einer Reise in den Columbianischen Anden. Hettner, in Petermann's Mitt. for 1888, pp. 104–112. Large folding map, plan of Bogotá, etc.

205. Beiträge zur geologie und petrographie der Kolumbianischen Anden. Hettner and Link. in Ztschr. Deutsch. Geol. Gesell.. 1888.

206. Beiträge zur Petrographie der Sierra Nevada de Santa Marta und der Sierra de Perijá in der Republik Colombia in Sudamerika. W. Bergt. in Mineral und Petrograph. Mitt. for 1889.

207. Die Sierra Nevada de Santa Marta und die Sierra de Perijá. Dr. W. Sievers, in Zeitschr. der Gesell. für Erdk. zu Berlin, 1888. Pp. 158, maps.

208. Erläuterungen zur geognostischen Karte der Sierra Nevada de Santa Marta. Sievers, in same for 1888. Map.

209. Die Kordillere von Mérida nebst Bemerkungen über das Karibische Gebirge, mit einer geologischen Karte und 15 profilen. Sievers. Vienna and Olmutz, 1888.

210. Der verfall des Staates Magdalena. Sievers. in Globus. 1888.

211. Die Floresta de la Santa Iglesia Catedral de la ciudad de Santa Marta. Sievers, in Globus for 1888.

212. Goajiro Halbinsel. J. Chaffanjon, in La Géographie for 1888.

213. Les mines d'or et d'argent de la Colombie. P. de Bruycker, in Bull. Soc. R. Géogr. Antwerp, 1888.

214. Le Sinou. Colombie. E. Patrouilleau, in Bull. Soc. Géogr. Comm. Bordeaux, 1888.

215. Atlas geográfico e histórico de la República de Colombia. MM. Paz and F. Perez. Paris, 1889. 20 maps, plans, views, etc.

216. Colombia: its past, present, and future. Reports from the Consuls of the U. S., 1889, pp. 98–112.

217. Reiseskizzen aus Kolumbien und Venezuela. Fr. Buchner. Munich, 1889.

218. Las estatuas del valle de San Augustín en la República de Colombia. J. Gutierrez de Alba, in Bol. Soc. Geogr. Madrid, 1889.

219. Die Goajiris Indianer. A. Sartori, in Mitt. Geogr. Gesell. Lubeck, 1889.

220. Reports and Recommendations of the International American Conference. Washington, 1890. (Colombia, pp. 122–127. Maps.)

221. Around and About South America: Twenty months of quest and query. Frank Vincent. New York, 1890. (Colombia, pp. 426–463.)

222. Notas de viaje: Colombia y Estados Unidos de América. S. Camacho Roldán. Bogotá, 1890. Pp. 6 + 900.

223. Avventure di una spedizione alla Colombia, per cura di M. Viglietti. Turin, 1890. Pp. 200.

224. Le miniere della Republica di Colombia. R. Ragnini, in Boll. Soc. Geogr. Ital., 1890, pp. 309–332.

225. La République de Colombie. H. Lennon, in Bull. Soc. Geogr., Antwerp, 1890, pp. 103–122.

226. Travels and Adventures of an Orchid Hunter: an account of Canoe and Camp Life in Colombia, while collecting Orchids in the Northern Andes. Albert Millican. London, 1891. One vol., pp. 15 + 222, 73 cuts, one colored plate.

227. Monumenti preistorici della Colombia; viaggio di G. M. Gutiérrez de Alba nella valle di S. Agostino. C. G. Toni, in L'Esplor. commerc., 1891, pp. 1–15.

228. Telegraphic determination of longitudes in Mexico, Central America, the West Indies, and on the north coast of South America. Norris and Laird. Bureau of Navigation, Washington, 1891.

229. Compagnie franco-belge des chemins de fer colombiens. R. Le Brun. Paris, 1891. Pp. 232.

230. Colombia. Bulletin No. 33, Bureau of the American Republics. Washington, January, 1892. One vol., pp. 138, 22 cuts, one map.

231. Nueva Geografía de Colombia. T. I. el territorio, el medio y la raya, Vergara. Bogotá, 1892.

232. Cartes Commerciales. 6me série. No. 10. Colombie et Equateur. F. Bianconi and E. Broc. Paris, 1892. Pp. 36, folding map.

233. Die Kordillere von Bogotá. Hettner, in Petermann's Mitt. Ergängzungsheft, 1892. Pp. 131, large map.

234. The Cordillera of Bogotá. Review of the above in P. R. G. S. of London for Dec., 1892, pp. 850–854.

235. La République de Colombie, géographie, histoire, etc.   R. Nuñez and H. Jalhay.   Brussels, 1893.   Pp. 259, map.

236. Die Anden des Westlichen Columbiens ; eine orographische skizze.   Hettner. in Petermann's Mitt. for 1893, pp. 129–136.

237. Reisen in Südamerika. Geologische studien in der Republik Colombia. III. Astronomische ortsbestimmungen, bearbeitet von Bruno Peter.   Reiss and Stübel.   Berlin, 1893.   Pp. 327.

238. Rio Hacha et les Indiens Goajires.   H. Candelier.   Paris, 1893.   Pp. 277, 41 cuts.

239. Coal and Petroleum in Colombia. Commercial information bulletin. Bureau of the American Republics.   Washington, 1893.

240. Encyclopædia Britannica, vol. vi., pp. 137–141.

There are articles upon Colombia in nearly all encyclopædias, and many short references in magazines and periodicals ; in particular, Das Ausland, Globus, etc., and in proceedings of the various geographical societies.

MAPS.

1. Atlas de los Estados Unidos de Colombia.   Codazzi.

2. Mapa de la provincia de Antióquia en la republica de Nueva Granada.   C. S. de Greiff.   Paris, 1857.

3. Large wall map of Colombia.   Thierry Brothers.   Paris, 1864.

4. Savanilla Harbor, Colombia.   1/36500, No. 925, Washington, 1885.

5. Cartagena Harbor.   1/36500, No. 978, Washington, 1886.

6. Colombie. Carte générale des chemins de fer projétés. 1/675000, Paris. 1886.

7. West Coast, Porto Bello.   1/36500, No. 958, Washington, 1887.

8. Parida and Palmque Anchorages.   1/73000, No. 1038.

9. Port Nuevo.   1/36500, No. 1039.

10. Bahia Honda.   1/36500, No. 1040, all Washington, 1887.

11. Map of River Sinú.   F. A. A. Simons, London, 1887.

12. Panama to Cape San Francisco.   1/975000, No. 1176, Washington, 1889.

13. Colombie. Port et mouillages. Port de Cispata, Port Careto, etc. No. 4633. Paris, Serv. Hydrogr., 1892.

14. Old Providence Island.   1/73000, No. 1372.

15. Santa Catalina Harbor.   1/18250, No. 1371.

16. Santa Marta Bay.   1/12150, No. 1378.

17. Gulf of Darien, Columbia Bay.   1/36500, No. 1405.

18. Gulf of San Blas, Mandinga Harbor.   1/36500, No. 1406.

19. Serrana Bank, South Cay Channel.   No. 1374.

20. San Miguel and Darien Harbor.   1/146000, No. 1410.

21. Chiri Chiri Bay.   1/18250, No. 1407, all Washington, 1893.

COLOMBIAN ZOÖLOGY.

1. The following is a partial list of the more extensive articles on the zoölogy of Colombia. It would require many pages to contain a complete list. Throughout the "Révue Zoologique," the "Magasin de Zoologie," the "Ibis," the Proceedings of the Zoölogical Society of London, and other similar publications are found many references to Colombian fauna. In the first two especially are found many descriptions of the discoveries of the naturalists Goudot and Delattre. The literature of Colombian humming-birds alone is voluminous : —

1. Fauna Cundinamarquesa. D. Jorge Tadeo Lozano. Mentioned by Pereira, but no date or locality given.

2. Notice sur quelques oiseaux de Carthagène, etc. Lafresnaye and D'Orbigny, in Rev. Zool., 1838, pp. 164–166.

3. Nouvelles espèces d'oiseaux mouches de Santa Fé de Bogotá. Boissonneau, in same, 1839, pp. 354–356.

4. Oiseaux nouveaux ou peu connus de Santa Fé de Bogotá. Boissonneau, in same, for 1840, pp. 2–8 and 66–71.

5. Coléoptères de Colombie, décrits par M. L. Reiche in same for 1842–43, total pp. 42.

6. Description de quelques oiseaux nouveaux de Colombie. Lafresnaye, in same for 1842, pp. 301, 302, and 333–336.

7. Insectes nouveaux observés sur les plateaux des Cordillères et dans les vallées chaudes de la Nouvelle-Grenade. Méneville and Goudot, in same for 1843, pp. 12–22.

8. Quelques oiseaux nouveaux ou peu connus de Colombie. Lafresnaye, in same for 1843, pp. 68–70 and 290–292.

9. Description de quelques coléoptères de la Nouvelle-Grenade. Méneville, in same for 1844, pp. 8–19.

10. Nouvelles espèces d'oiseaux de Colombie. Lafresnaye, in same for 1844, pp. 80–83.

11. Coup d'œil sur l'ornithologie de la Colombie. Lafresnaye, in same for 1845, pp. 113–119.

12. Description de quelques mammifères Américains. Pucheran, in same for 1845, pp. 335–337.

13. Sur quelques nouvelles espèces d'oiseaux de Colombie. Lafresnaye, in same for 1846, pp. 206–209.

14. Description de quinze espèces nouvelles de trochilidées. Delattre and Bourcier, in same for 1846, pp. 305–312.

15. Description de vingt espèces d'oiseaux mouches. Bourcier, in Annal. de la Soc. Royal d'agricult., etc., de Lyons for 1846.

16. Sur le ramphocelus icteronotus du Prince Bonaparte. Lafresnaye, in Rev. Zool. for 1846, pp. 365–370.

17. Quelques oiseaux nouveaux de Bolivie et de Nouvelle-Grenade. Lafresnaye, in same for 1847, pp. 65–79.

18. Sur l'espèce de rhamphocèle à plumage variable rapporté de la Nouvelle-Grenade. etc. La Fresnaye, in same for 1847, pp. 215–219.

19. On the birds received in collections from Santa Fe di Bogotá. Sclater, in P. Z. S. for July. 1855. Pp. 36. 435 species enumerated. This was also published as a separate paper.

20. On some additional species of birds received in collections from Bogotá. Sclater, in same for 1856, pp. 25–31. 4 colored plates.

21. Further additions to the list of birds received in collections from Bogotá. Sclater, in same for 1857. pp. 15–20. 52 species.

22. Catalogue of birds collected during a survey of a route for a ship canal across the Isthmus of Darien. etc. Cassin, in Proc. Ac. Nat. Sci. of Phil. for 1860, pp. 132–144 and 188–197. 144 species.

23. Note sur les trochilidées de la Nouvelle Grenade. De Geofroy. Bogotá and London. 1861.

24. Catalogue of a collection of birds made in New Grenada. etc. G. N. Lawrence. in Annals Lyc. Nat. Hist. N. Y. for 1861, 62, 63. 4 parts, total pp. 67.

25. Descriptions of six new species of birds from the Isthmus of Panama. Lawrence, in the Ibis for 1862, pp. 10–13.

26. Descriptions of eight new species of birds from the Isthmus of Panama. Lawrence, in same for 1863, pp. 181–184.

27. Notes on a collection of birds from the Isthmus of Panama. Sclater and Salvin, in P. Z. S. for 1864, pp. 342–373. One colored pl.

28. Description of eight new species of birds from Veragua. Salvin. in same for 1866, pp. 67–76. Two colored plates.

29. On some collections of birds from Veragua. Salvin. in same for 1867, pp. 129–161. Colored plate.

30. On some collections of birds from Veragua. Salvin, in same for 1870. pp. 175–219. Map.

31. Notes on some birds of the United States of Columbia. C. W. Wyatt, in same for 1871, pp. 113–131, 319–335, 373–384. Map.

32. On a collection of birds from the Sierra Nevada of Santa Marta. Columbia. Salvin. in same for 1879. pp. 196–206.

33. On the birds collected by the late Mr. T. K. Salmon in the state of Antióquia, United States of Colombia. Sclater and Salvin, in P. Z. S. for 1879. pp. 486–550. Map and three colored plates.

34. On the birds of the Sierra Nevada of Santa Marta, Columbia. Salvin and Godman, in Ibis for 1880. pp. 114–125, 169–178. 3 plates.

35. Untersuchungen über die Vögel der Umgegend von Bucaramanga in Neu Granada. Von Berlepsch, in Journal für Ornithologie for 1884, pp. 273–320.

36. On some interesting additions to the avifauna of Bucaramanga, U. S. of Colombia. Von Berlepsch, in Ibis for 1886, pp. 53–57. Plate.

WORKS ON CURAÇAO.

1. See No. 17 in list of works on Colombia. Curaçao (Curassou) is in fifth vol. p. 425 et seq.

2. Description of a new species of humming-bird from the island of Curaçao. Lawrence, in Annals Lyc. Nat. Hist. N. Y., p. 13.

3. Eenige West Indischen Kolonien na de emancipatie. Fraissinet. Amsterdam, 1879. Pp. 43.

4. Curaçao. J. Kuyper, in Tijdschr. aardrijksk. genootsch. 1882.

5. The Aruba and the Papiamento Jargon. Gatchet. American Phil. Soc. Philadelphia, 1884.

6. On a collection of birds made by Messrs. Benedict and Nye, etc., Island of Curaçao, Venezuela [*sic*]. Ridgway, in Proc. U. S. Nat. Museum for 1884, pp. 173–177.

7. Die Niederlandische Expedition nach den Westindischen Inseln und Surinam, 1884–85. Martin, in Tijd. aard. genoot. Amsterdam, 1885.

8. Reise nach den Niederländisch Westindischen Besitzungen. K. Martin, in Rev. Colon. Internat. 1885.

9. Nederl. Westindische Expeditie. W. F. R. Suringar, in Tijd. Nederl. aard. genoot. Amsterdam, 1886.

10. Overhet geolog. verband tusschen de Westindische eilanden. Dr. Molengraff, in same for 1887.

11. Geolog. Kaarten van Curaçao, Aruba en Bonaire. C. M. Kan, in same for 1887. Three maps.

12. Note sur la latitude de Curaçao et sur les longitudes de Laguayra, Puerto Cabello, Curaçao et Sainte Marthe. M. Aubry, in Annales Hydrogr. Paris, 1887.

13. Bericht über eine Reise nach Niederländisch Westindien und darauf gegründete Studien I. Land und Leute. Martin. Leiden, 1887.

14. West-indische Skizzen. Martin. Leiden, 1887. Map, pp. 186.

15. Beiträge zur Geologie von Niederländisch Westindien. Martin. Leiden. 1887–89.

16. Les possessions néerlandaises dans les Antilles. T. C. L. Wijnmalen. Amsterdam, 1888.

17. Statistisch overzicht van Ned. West Indie. Same in Bijdr. Stat. Inst. 1888.

18. Geologische Studien über Niederländische-Westindien. Martin. Leiden, 1888.

19. Die Vogel der Insel Curaçao nach einer von Herrn Cand. Theol. Ernst Peters daselbst angelegten sammlung. Von Berlepsch, in Journ. für Ornithol., Jan. 1892. Pp. 62.

20. Under the Southern Cross: a guide to the sanitariums and other charming places in the West Indies and Spanish Main. W. F. Hutchinson. Providence, 1892.

21. Ernst Hartert. Remarks on some birds from Curaçao, in Bull. Brit. Ornithol. Club. Ibis. Jan., 1893.

22. On the birds of the Islands of Aruba, Curaçao, and Bonaire.   Hartert, in Ibis for July, 1893.   Pp. 50, map, colored plate.

23. L'ile de Curaçao.   G. Verschuur, in Tour du Monde for 1893, pp. 81–96. 12 cuts, map.

24. Encyclopædia Britannica, vol. vi. p. 626.

25. Kaart von het Eiland Curaçao, etc.   Amsterdam, 1886, 20 by 24 ins.

26. Santa Ana Harbor.   No. 1049, 16 by 19 ins.   Washington, 1890.

27. Spanish Water, Spanish Haven, and Caracas Bay.   No. 1245, Washington, 1891.